NUCLE.

"Jack Spencer tells it the way it is and needs to be to make nuclear energy an affordable choice for America. Read this thoughtful and provocative book written by one of America's great conservative thinkers."

WILLIAM MARTIN, former Deputy Energy Secretary; Chair, Nuclear Energy Advisory Committee

"Members of Congress agree on very few things. Bipartisanship is extremely rare in politics today. Confronting and containing an expansive China is one policy where there is widespread consensus. The other area of agreement, surprisingly, is the indispensable role of nuclear energy in the twenty-first century. In his thought-provoking new book, *Nuclear Revolution*, Jack Spencer chronicles this transformation in energy politics and policy, and offers an alternative free-market approach to realizing nuclear's full potential. Driven by climate concerns and energy security risks globally—and with AI now surging energy demand—nuclear is the one clean, secure, and abundant energy source. Spencer's candid assessment of where we have over-subsidized, overregulated, and overblown expectations is an important wake-up call and a timely prescription for rethinking energy policy in the US and globally."

SCOTT CAMPBELL, President, Howard Baker Forum; Director, U.S.-Japan Roundtable on Nuclear Energy; Senior Strategic Advisor, Baker Donelson

"Jack Spencer makes a convincing and entertaining argument that nuclear power, while an essential part of the American energy portfolio, is not a cure-all for whatever risk one might assign to climate change, nor is it likely to prosper from increased federal 'support.' Along the way, he cuts holes in the politicians, special interest groups, and bureaucrats who have decided to make choices for the rest of us with respect to everything from cars to uranium fuel, and who have, in his words, 'created entire narratives about energy scarcity and environmental degradation to justify their power grabs.' Buy the book, get some popcorn, and enjoy the ride."

MIKE McKENNA, President, MWR Strategies, and contributing editor at *The Washington Times*

"In *Nuclear Revolution*, Jack Spencer lays out the conservative's approach to igniting a resurgence of nuclear energy in the US. You may not agree with his take, but he brings receipts to the discussion and provides a point of view that can't be ignored."

CRIAG PIERCY, Executive Director and CEO, American Nuclear Society

"Jack Spencer is one of the most creative and innovative thinkers on nuclear energy policy. Nuclear energy's success in America hinges on bold, imaginative policy reforms; this book provides that road map. *Nuclear Revolution* should be required reading for any member of Congress and anyone who wants nuclear power to flourish."

NICK LORIS, Executive Vice President, C3 Solutions

"The economic and environmental advantages of nuclear power are clear. Although some argue that nuclear energy is too expensive, the majority of the costs are tied to construction. As Jack Spencer points out in *Nuclear Revolution*, reactors don't need to take decades to complete. Streamlining the government approval process can make nuclear energy cost-effective, and Spencer brings an interesting perspective on how to achieve that."

FLORENCE LOWE-LEE, President and Founder, Global America Business Initiative

"President Ronald Reagan said, 'Government's view of the economy could be summed up in a few short phrases: If it moves, tax it. If it keeps moving, regulate it. And if it stops moving, subsidize it.' As Jack Spencer points out in *Nuclear Revolution*, the same holds true for the government's view of this vital component of our nation's energy mix. Policy failures, fearmongering, and just plain bad decision-making has veered the nuclear power industry way off track. Spencer breaks these down one by one and corrects the many myths about nuclear technology along the way. No one knows what the future holds for nuclear power, but this book provides a policy road map to at least give it a greater opportunity to compete for the reliable electrons necessary to power our future."

THOMAS PYLE, President, Institute for Energy Research

NUCLEAR REVOLUTION

POWERING THE NEXT GENERATION

JACK SPENCER

OPTIMUM PUBLISHING INTERNATIONAL
LONDON | MONTRÉAL | TORONTO

Nuclear Revolution, Powering the Next Generation
© Ottawa and Washington 2024, Optimum Publishing International

First Edition © 2024 Optimum Publishing International and Jack Spencer
Published by Optimum Publishing International

All rights reserved. No part of this publication may be reproduced in any form or by any means whatsoever or stored in a database without permission in writing from the publisher, except by a reviewer who may quote passages of customary brevity in review.

Library and Archives Canada, Cataloguing in publication
US Copyright Case #1-14394982205

Jack Spencer
ISBN: 978-0-88890-357-0 (Paperback, first edition)
Digital version of the book is also available ISBN: 978-0-88890-358-7

Nuclear Revolution, Powering the Next Generation
I. Title

Canadian edition is printed and bound in Canada
US edition is printed and bound in the USA

For information on rights or any submissions, please write to:

Optimum Publishing International
Toronto, ON
Dean Baxendale, President
Website: www.optimumpublishinginternational.com
Twitter: @opibooks

CONTENTS

FOREWORD		vii
INTRODUCTION		xi
1.	WHY NUCLEAR ENERGY?	1
2.	SECURITY THROUGH AFFORDABLE, CLEAN, SAFE ENERGY	18
3.	PLAGUED BY MYTHS	31
4.	THREE MILE ISLAND, CHERNOBYL, AND FUKUSHIMA DAIICHI	44
5.	THE BROKEN POWER PLANT FALLACY	55
6.	AN INDUSTRY GASLIGHTED	68
7.	GOVERNMENT CAN'T SUBSIDIZE AN INDUSTRY INTO SUCCESS	81
8.	THE PROMISE, CHALLENGE, AND OPPORTUNITY OF ADVANCED REACTORS	99
9.	FUELING THE TWENTY-FIRST CENTURY	112
10.	THE ATOMIC OPPORTUNITY OF NUCLEAR WASTE	130
11.	A POLICY REVOLUTION	145
ACKNOWLEDGMENTS		165
ENDNOTES		169
INDEX		199

FOREWORD

by Stephen Moore

I got to know Jack Spencer over a decade ago when I returned to the Heritage Foundation, where we both worked on economic and energy policy. At the time, I was finishing up work on *Fueling Freedom: Exposing the Mad War on Energy*, a book I co-authored with Kathleen Hartnett White, who we recently and very regrettably lost.

I can't count how many times Jack and I sat in his or my office and discussed for hours this or that policy, how to defend against the latest attacks on free enterprise, or the insanity of the government's most recent efforts to plan all our lives. Almost without fail, regardless of how the conversation started, we would somehow get to nuclear energy. We discussed at length how nuclear energy could and would make a comeback—and boy, do we look right about that now, given the new sparks of life in the industry.

I have spent a career arguing for policies that result in greater economic growth. We know that a low tax and low regulatory environment is essential to unleashing prosperity for all Americans. But prosperity is not born only of these principles of economic freedom. Prosperity also requires energy. It is no surprise that when energy is abundant, the economy takes off. And the opposite is also true: energy scarcity results in economic trouble.

Despite the scare tactics of the 1970s—anyone remember the anti-nukes Hollywood blockbuster *The China Syndrome*, starring Jane Fonda?—we now know that nuclear power is safe, clean, and capable of providing immense amounts of energy. American companies have built over a hundred commercial reactors over the years, so as an economist, I know that nuclear had to make economic sense at some point. And I know that there's a reason why power-hungry nations around the world continue building nuclear plants today.

What I didn't understand is why the United States stopped building nuclear power plants. Of course, I understood the general narrative that low-cost alternatives like natural gas create stiff competition for nuclear energy, that decades of antinuclear propaganda have soured the public on the technology, and that regulation artificially inflates cost. But those reasons all seemed surmountable if we tried, yet we appeared powerless to do anything real about them.

Jack, I soon learned, had become one of the foremost experts on nuclear power, and he was for it long before it became cool to be for it.

But Jack isn't a shill for the industry. There are public policy challenges with respect to safety, insurance issues, nuclear waste concerns, and so on, and these need to be thoughtfully addressed if nuclear power is to grow as a major energy source for the United States and the rest of the world.

Jack also rejects subsidies across the board. Like I do, he rejects all energy subsidies. He wants nuclear to succeed or fail on its own merits.

Here is a secret of the policy world: Many of us toil in relative obscurity in our issue area for years. We think. We write. We develop ideas. People often don't really understand the relevance of what we do because it lies outside of the political discussion of the day. But we continue our work and wait for our time to come. One day something happens that brings our issue to the headlines, and those years of work pay off.

This book could not be more expertly timed. Nuclear energy is ready for liftoff with Google, Amazon, and Microsoft signing on to big nuclear power deals just in recent months.

I have warned for years that the so-called green energy agenda is leading the nation to a point of energy poverty. Given its technological prowess and vast natural resources, the United States should be enjoying an era of unprecedented energy abundance. Yet we find ourselves on the edge of perpetual energy scarcity—a situation that will only get worse given the massive energy demands of the next generation of technology, particularly AI.

Of course, scarcity is not necessarily a bad thing. It signals to the broader economy that we need to find new alternatives, and it drives innovation and competition. But when that scarcity is imposed by government and exacerbated by drastic limitations on what alternatives can be brought to market, we are headed for trouble. That would be bad enough, but the situation is so much worse. Our regulatory environment makes building and creating almost impossible, our tax policies disincentivize investment, and our infinitely complex system of subsidies and mandates obscures supply-and-demand signals beyond recognition.

Free enterprise can withstand some government intervention, but what Washington has done to the energy economy is unsustainable. It has crippled nuclear energy in particular. This at a time when our electric power grid is vulnerable to blackouts and brownouts.

The idea that energy shortages are coming is not theoretical. The organizations whose job is to monitor electricity supplies have begun warning that the nation is facing massive shortages. If we don't do something now, the United States will have wide-scale blackouts soon. The lights won't turn on, and we won't be able to heat and cool our homes when we need to most. America's electricity grid is literally breaking down.

I'm not the only one realizing this. It seems that the world has realized it too. In fact, one of the big announcements to come out of last year's global warming conference was a commitment to triple nuclear energy. Russia and China have shifted their industries into high gear. Japan is working to bring its reactors back online. And South Korea continues to develop its commercial nuclear industry as well. Enthusiasm in the United

States is just as high, with legislation that allots huge quantities of taxpayer money to nuclear energy, and that begins to address the 1970s regulations hamstringing our twenty-first-century power demands.

But this is also the problem—and the reason why this book is so important. As a student of economics, I understand that any industry built on government support within a heavily regulated environment will never succeed over the long term. We need nuclear to succeed and become a major part of our energy future.

It will happen, and this book is the road map that explains how. If we want to keep the lights on, keep our schools and hospitals and factories open, keep our superpower status, we'd better pay attention to this book.

INTRODUCTION

Nuclear energy is having a moment.

Republicans support nuclear energy because it provides national and energy security benefits. Democrats support it because it's the only credible pathway to meeting their climate objectives. The once mighty and influential antinuclear movement has virtually no seat at the table, and the industry itself is no longer defined by disasters like Three Mile Island and Chernobyl. Even the Fukushima accident in Japan in 2011 is ancient history to younger Americans.

Hollow scaremongering about the dangers of nuclear energy has been supplanted by decades of safe operation, reliability, and familiarity. Communities nationwide have not only lived with nuclear power but thrived with it. The debate over the integrity of nuclear energy is over. Experience demonstrates that it provides good jobs, clean energy, and dependable power.

As the late conservative writer Andrew Breitbart famously said, "Politics is downstream from culture," and that is nowhere truer than when it comes to nuclear energy. When the fearful cultural aura around nuclear power started to fade, the politics also began to shift. And now, in a deeply divided America, this is one issue on which liberals and conservatives—Democrats and Republicans—largely agree, at least in principle.

Today, a clear national consensus is emerging that nuclear energy is important to the future of the country, and that our political leaders need to act to bring about that future. None of this would have happened if American culture had not shifted in a pronuclear direction. The politics is following.

But public acceptance alone will not result in a vibrant, competitive, and sustainable industry. Whether we like it or not, economics matter. No industry can succeed if it's not economically competitive—and this must be the focus of the debate over nuclear energy in today's discourse. While the antinuclear movement has been largely cast aside, its legacy looms in every corner of the American nuclear energy industry and the halls of power that affect its economic viability. Skepticism sown through decades of misinformation around nuclear power from those who opposed it continues to inform how the industry is regulated.

Indeed, this skepticism has resulted in a regulatory approach that imposes unnecessary costs, has hindered innovation, and has left our political and industry leaders fearful of championing the sorts of reforms that most would recognize are necessary and legitimate. While nuclear does currently enjoy broad public and bipartisan political support, little is being done to reform the underlying policies and regulations that got the industry to the suboptimal position it is in today.

To be clear, this is not to criticize the nuclear industry. That it even exists at all is testament to its value and resiliency. For nearly six decades, America's nuclear companies have had to deal with whipsaw policies that try to push the industry forward at one moment and shut it down in the next. America's incoherent nuclear energy policies can often do both simultaneously. So while this book will argue against subsidies for nuclear energy (or any other type of energy), it will not criticize firms for accepting those subsidies. No one in their right mind would invest their own money in an undertaking so wrought in political risk if the government, which largely created that risk in the first place, didn't do the same.

People on both sides of the debate will argue that nuclear energy is just too expensive to move forward. Those who support it will say that subsidies are needed to make the economics work and help validate new technologies. Those who oppose it will say that no amount of subsidy will create a viable industry.

The truth is that no one knows what the future of nuclear energy looks like. But we do know that building a nuclear power plant used to be far more affordable. We also know that people used to be willing to invest in nuclear power even without direct taxpayer subsidies—or at least with many fewer of them. Earlier nuclear power plants were built with very little direct government support. And we know that in those early days, those tasked with moving nuclear energy from the lab into the market knew full well that the private sector, not government, was the key to the industry's commercial success.

There is no question that government had a role in nuclear energy's inception, and that it will continue to have a role into the future. But what should that role be?

I've had the opportunity over my career to visit countless commercial nuclear facilities and get to know many individuals in the industry, and I can say without reservation that they are the finest, smartest, most innovative people our country has to offer. I have absolute faith in their ability to push nuclear energy forward toward successes beyond our imaginings. But they need to be free to do it—and that's what this book is about. It rejects the idea that government can lead the American nuclear industry into a prosperous future. It is not about how to model the success of the French, the Russians, or the Chinese. This book is about how to build a uniquely American industry—one that is driven by competition, innovation, and entrepreneurship. It should be animated by the American spirit, molded by free enterprise, and delivered by the private sector. What we need in American nuclear power is not a renaissance but a revolution.

To achieve that revolution, we first need a revolution in policy and regulation. The approach proposed in the pages that follow, however, does not advocate burning down the existing structures (for the most part). Rather, it offers alternatives to them. It argues that subsidies are counterproductive and should be eliminated. This will level the playing field for all energy sources. It then recommends a series of reforms that properly align incentives and responsibilities, allowing all energy industries to move forward according to their own merit as determined by the people and businesses that ultimately use and pay for their products.

If lawmakers truly believe what they say about the promise and potential of nuclear technology, then tinkering around the edges won't do. The time has come for a policy revolution.

CHAPTER 1
WHY NUCLEAR ENERGY?

If one thing is as true as death and taxes, it's that modern industrial nations require a lot of energy, and that energy demand grows over time. This is not something that we as a society can work around—it is an absolute necessity. Countries that have access to energy are better off by almost any standard than those that do not. It is needed for life—or at least modern life. A life where our children are better off than we were, and where each generation can expect to live a little longer than the one before it. As Alex Epstein writes, in his book *Fossil Future*, "The essential value of energy and machines to human flourishing is that they *amplify* and *expand* our natural meager productive ability."[1] Without abundant energy, there is no work-life balance. There is just work.

History shows that a growing standard of living is always accompanied by increased energy demands. No amount of political hand-wringing will change this simple fact. Nor will bureaucratic schemes for energy efficiency, energy transitions, or whatever regulatory framework bureaucrats want to impose. As politicians make energy more expensive and less available, our lives will become more difficult.

Just look at how human well-being—as measured by wealth, health, and life expectancy—has skyrocketed since the Industrial Revolution, which was fueled by access to energy.[2]

Let's break it down. Prior to 1700, per capita gross domestic product (the sum value of all goods and services produced within a nation's borders) was stuck at around $955 (in 2019 dollars) per year in the West.[3] Today, the average North American can expect a per capita GDP of around $66,935 (also in 2019 dollars).[4]

While historians and economists may debate at the margins, most can agree that two things were key to this astronomical rise in economic production. First was the spread of free enterprise (thank you, Adam Smith), and second was the broad availability of affordable, scalable, and efficient energy (thank you, hydrocarbons and uranium).

Today, nations with low energy consumption (defined as 500 kilowatt-hours [kWh] per capita or less) have economies characterized by subsistence-level output, with annual per capita incomes of roughly $1,000. In contrast, countries that consume around 100,000 kWh per person tend to have much higher per capita incomes, allowing their populations access to basic modern comforts such as clean running water, refrigeration, and air-conditioning.[5]

There are more than 335 million Americans today. Had the nation stayed on the same GDP trajectory that humans had been on for hundreds of years prior to the energy revolution, we would have a GDP today of around $317 billion. Subtract that from 2022's actual GDP of approximately $22.24 trillion and you get $21.92 trillion![6] In other words, we generate and consume nearly seventy times more goods and services with affordable, abundant energy than we would have without it.

OH, THE ENVIRONMENT

It's clear that energy and human well-being go hand in hand. Despite that, politicians seem to be doing everything they can mess up a good thing. They are trying to impose drastic limits on hydrocarbons like gas, oil, and coal, and almost no politicians are advocating for the sorts of revolutionary policy reforms that would allow nuclear energy to fully reach its potential.

Of course, they give their reasons. They say that hydrocarbons are destroying the environment. We should always strive to ensure a healthy environment, but the data does not back the notion that energy use is leading to environmental degradation. In fact, according to the Environmental Protection Agency (EPA), US air quality has been improving for decades. EPA reports that since 1990, air pollutant concentration levels have declined significantly nationwide:[7]

- carbon monoxide (CO) eight-hour: down 79%
- lead (Pb) three-month average: down 85% (from 2010)
- nitrogen dioxide (NO_2) annual: down 61%
- nitrogen dioxide (NO_2) one-hour: down 54%
- ozone (O_3) eight-hour: down 21%
- particulate matter 10 microns (PM_{10}) twenty-four-hour: down 32%
- particulate matter 2.5 microns ($PM_{2.5}$) annual: down 37% (from 2000)
- particulate matter 2.5 microns ($PM_{2.5}$) twenty-four-hour: down 33% (from 2000)
- sulfur dioxide (SO_2) one-hour: down 91%

By almost any measure, America's air quality is clean and getting cleaner. So the argument that a forced energy transition is justified for traditional environmental reasons is unpersuasive.

Next, hydrocarbon opponents fall back on climate change. The truth here is that there has been a dramatic decline in climate-related deaths over the past one hundred years. In fact, these deaths are down a staggering 92 percent since the 1920s, when the statistic was first recorded.[8]

Then they say we need to stop using hydrocarbons now to protect the future. Even if we put aside the real debate over the human effect on climate and accept for the purposes of this discussion the underlying assumptions that drive the left's energy policy, capping hydrocarbon use would have virtually no impact on climate.[9] In fact, eliminating *all* carbon dioxide emissions in the United States would have almost no environmental benefit.

The Heritage Foundation's chief statistician, Kevin Dayaratna, has investigated this using the same models used by government agencies. He found that eliminating all US greenhouse gas emissions would reduce temperatures by around 0.2 degrees Celsius by 2100.[10] Eliminating all emissions from OECD countries would result in a reduction of only 0.5 degrees.[11] For context, the Paris Accords aim to limit the increase in global temperatures to 1.5 degrees.

There should be no debate. Unless we are fine with our children being worse off than we are now, we need affordable, reliable power, and hydrocarbons are a perfectly good source of that.

Indeed, the US government predicts that by 2050, electricity demand will increase by up to 22 percent among residential customers and up to 38 percent among industrial users.[12] At the same time, the North American Electric Reliability Corporation (better known as NERC), whose mission is to ensure the reliability of the electricity grid, is warning that nine of the thirteen assessment zones in America are at elevated or high risk of energy shortfalls by 2028.[13] The Midcontinent Independent System Operator (MISO) region, which powers much of the Midwest, is one of the high-risk zones (the other is SERC-Central, which powers all of Tennessee and parts of Missouri, Kentucky, and other neighboring states). MISO is predicted to face a 4.7-gigawatt energy shortfall—enough to power more than three million homes—due to increased demand and retirement of existing power plants.

What about the rest of the world?

By 2050, global electricity demand will be even higher, with the US government projecting increases of as much as 75 percent.[14] And if the left's energy policies prevail, the shortfalls will be even worse.

Many people have many thoughts about how to produce and consume the power that will be needed, and they express them through policy preferences and market decision-making. Some people want cheap power.

Others want CO_2-free power. Still others put a priority on domestically produced power. The preferences are virtually unlimited, but almost everyone can agree that we want power that is abundant, reliable, clean, and affordable. The marketplace is where all of this should be sorted out.

Some people, though—like politicians, special interest groups, and bureaucrats at all levels of government—aren't satisfied with simply expressing their preferences through consumer choice. They have decided to use their political and legal authority to make many of these choices for the rest of us. Indeed, they have created entire narratives about energy scarcity and environmental degradation to justify their power grabs.

How did we get here?

The foundation for our current approach to energy policy was laid during the energy debates of the 1970s. Back then, America was facing severe energy shortages just at the time the environmental movement was beginning to take shape. This produced a political environment that was receptive to the idea of a heavy-handed energy industrial policy.

In fact, entire institutions, including the Department of Energy, were created for the specific purpose of making energy production and consumption decisions for us. Even today, Washington bureaucrats, politicians, and special interests are quick to defend the Department of Energy and increase its power, but the underlying justifications for its existence are completely outdated.

Section 101 of the Department of Energy Organization Act of 1977, which established the DOE, outlines why the department was needed. Among the justifications are that "(1) the United States faces an increasing shortage of nonrenewable energy resources; (2) this energy shortage and our increasing dependence on foreign energy supplies present a serious threat to the national security of the United States and to the health, safety and welfare of its citizens; (3) a strong national energy program is needed to meet the present and future energy needs of the Nation consistent with overall national economic, environmental and social goals."[15]

During the 1970s energy crisis, this was all arguably accurate, but none of it remains true today. Indeed, the United States now enjoys unprecedented energy abundance. According to the US Energy Information Administration (EIA), in 2020, the United States held over 373 billion barrels of technically recoverable crude oil reserves[16]—that's equivalent to a fifty-year supply. The agency also estimated that the US held 2,925.8 trillion cubic feet of technically recoverable natural gas. At current consumption levels, that equates to nearly one hundred years of supply. What's more, there are always new discoveries that would expand this supply over time, as demonstrated by the growing availability of unconventional sources like oil shale. According to EIA, the US currently holds 195.4 billion barrels of crude oil and 1,712.9 trillion cubic feet of natural gas in unconventional reserves.

According to the Institute for Energy Research (IER), combined conventional and unconventional natural gas reserves would provide well over a thousand years of energy supply at 2022 consumption levels.[17] Then there is coal. According to IER's most recent study on the subject, the United States has nearly five hundred years of proven coal reserves and more than nine hundred years of technically recoverable supply (based on 2022 consumption rates).[18]

These energy resources do not even account for uranium deposits, which are widely spread throughout the US, or for advancements in energy-generating technologies that we may not even have discovered yet.

This abundance is why the United States is largely energy independent.[19] While there was certainly a time when America's dependence on foreign energy was a concern—to the Organization Act's second point—those days are long gone. That independence undercuts the act's third point because it was achieved not through some national energy program but through private markets discovering new and innovative ways to produce energy, like hydraulic fracturing.[20]

That energy access is a problem for any American today is purely a product of bad policy that makes developing technologies and energy resources more difficult than it would otherwise be. Despite this, most Americans, including representatives of both major political parties, essentially accept government's role in energy production and consumption.

While free energy markets would be the optimal system to match consumer preference with the types and costs of energy produced, the truth is that most existing markets are heavily regulated and politicized and produce suboptimal outcomes (if the optimal outcome is defined as that which suits the preferences of the most people).

GOVERNMENT INTERVENTION LEAVES INDUSTRY UNPREPARED

One of the great challenges of an approach that is defined by government rules and regulations is that when faced with adversity, energy production and distribution systems are unprepared to respond as efficiently and reliably as they could. That's because firms must build their businesses around how to accommodate political preferences, not how best to meet customer demand—and these goals often contradict.

To understand what's driving energy policy today, consider Washington's efforts to electrify the auto industry. Consumers clearly do not want to replace their gasoline-powered vehicles with electric ones, yet regulations are being imposed that essentially mandate the transition. This forces the auto industry to move away from what customers want and toward what the government demands. This will almost certainly leave America's auto industry weaker as those contradictory objectives resolve in coming years. The broader energy industry is no different. Companies can't focus on providing consumers with affordable and reliable energy because they are preoccupied by trying to account for the ever-growing reams of rules and regulations being imposed by all levels of government. This necessarily will result in an industry that is not optimized to meet consumer demand, which will leave it weaker in the future.

Few politicians, bureaucrats, or special interest groups ever acknowledge that this suboptimal performance is the direct result of their interventions; instead, they routinely blame it on the industry itself or even the lack of rules and regulations. This narrative emboldens policymakers to attempt additional interventions to manage energy markets to accommodate the varied interests of their many stakeholders.

This dynamic is playing out in real time and dramatic fashion. In an alleged effort to provide Americans with secure, clean, abundant, and affordable power, policymakers have created such a complex and distorted energy economy that producers are simply unable to meet all these needs, even during the best of times. This is why gasoline prices have stabilized at a significantly higher level than they were just a few years ago, and why NERC is predicting shortfalls in electricity production.

In these challenging times of global pandemics, geopolitical instability, and economic disruption, the inefficiencies of America's suboptimal energy infrastructure are exposed by problems that policymakers have created. Put simply, America's energy economy will not efficiently respond to either supply or demand signals. In other words, the system won't work for either energy producers or consumers. Fortunately, the American economy has not yet felt the full brunt of decades of an anti-energy agenda.

Some other countries have not been so lucky.

Germany is a good example of what happens when arrogant politicians think they can better manage complex systems than a marketplace of producers and consumers. The country saw energy prices skyrocket following Russia's invasion of Ukraine, and the impact on German industry was felt almost immediately, with companies paying around 40 percent more following the invasion than before.[21] The German Chambers of Commerce and Industry reported that industry pessimism about the economy was as severe in the wake of the invasion as it was during the 2008 financial crisis and the initial Covid lockdowns.[22] Seventy-eight percent of companies surveyed identified rising energy and raw material prices as one of the biggest risks they faced and said they expected that their businesses would deteriorate. Another survey found that nearly 25 percent of Germany's small- and medium-sized businesses were considering or actively relocating parts of their operations to other countries.[23]

This was completely predictable, and here is why.

Germany is attempting to reach so-called net-zero carbon dioxide emissions by 2045. To achieve this, all electricity must be produced using renewables by 2035, and all coal must be phased out. Although legislation sets 2038 as the latest date for ending coal power, market forces are likely to accelerate this timeline. Rising costs due to the EU Emissions Trading System are expected to make coal economically unviable by around 2030, effectively phasing it out earlier than mandated. And for no good reason at all, the Germans are phasing out nuclear energy as well.

None of this made sense from an economic, environmental, or logical standpoint, but the country was going along with the program, pretending that its energy policy was doing just fine. Then Russia invaded Ukraine, and everything and nothing changed.

Though Germany was attempting to impose a transition to renewable energy when Russia invaded, it still relied on hydrocarbons for over three-quarters of its energy use.[24] Within that, 24.1 percent of its energy supply came from natural gas, and 55 percent of that came from Russia. Germany was directly dependent on Russia for its energy.

This was a critical lifeline to the economy. So long as the Russian energy taps remained open, German politicians and bureaucrats could impose almost any irrational energy policy they wanted and essentially moderate the negative impact of bad policy by importing more Russian fuel. If by some stroke of God, Germany's quest for net zero worked, it could ramp down Russian imports over time. The more likely scenario, however, was that their policies would fail, and Russian gas imports would be there to save the day. The problem with this approach became obvious in April 2022.

There is no doubt that geopolitical turmoil, such as Russia's war on Ukraine, can spark price spikes. But policies that restrict gas, oil, and coal development—in both the United States and Europe—exacerbate and extend price hikes and can create full-blown energy crises. That's what happened in Germany.

German politicians disregarded their citizens' demands for affordable and abundant energy, opting to discourage traditional power generation in favor of their preferred "green" alternatives. As market power shifts away from consumers and toward governments, private investment often follows. The result in Germany was underinvestment in gas, oil, coal, and nuclear, with the money flowing instead to less reliable, more expensive sources, like wind and solar. Officials realized they would face a voter revolt if citizens had to pay the full price of this policy, so they tried to ease the pain of transition by turning to what they believed would be a reliable source of energy: Russian oil and gas. This was manageable until Putin's natural gas supplies were no longer available.

Germany's problems were avoidable. The European Union has substantial energy reserves: nearly 14 billion tonnes in proven coal reserves alone.[25] That's more than eighty-five times the EU's total coal consumption last year (160 million tonnes).[26] Europe also has over 7 trillion cubic meters of natural gas resources.[27] EU natural gas consumption in 2021 was approximately 397 billion cubic meters.[28] And Europe is thought to have even more recoverable shale gas than the US.[29] That's enough coal and natural gas to power European homes and businesses for many decades.

In addition, Europe has underutilized nuclear power. The continent has over a hundred reactors providing about a quarter of its electricity.[30] But half of those reactors are in France, which gets around 70 percent of its electricity from nuclear power. Half of the EU's member states have no nuclear reactors at all.

Europe's energy woes have little to do with access to affordable, reliable, and clean energy and everything to do with pushing unrealistic policy choices. And families and businesses are paying a steep price for those poor decisions.

Unfortunately, it doesn't look like things are changing much in Germany. Rather than recognize that current policy fell far short of what the people deserve, politicians are doubling down on their overall approach and intentions. The country remains committed to reaching net zero, although there is some acknowledgment that the timeline may need to be expanded.[31] Of course, that was always going to be the case because the timeline is impossible to achieve.

Now, rather than institute market reforms that would allow prices to stabilize through the organic interaction of supply and demand and encourage investment to flow toward what makes the most economic sense, German politicians have decided to use subsidies and other taxpayer-funded schemes to mitigate the real-world impacts of their energy policy.[32]

The approach will never work. Politicians and bureaucrats simply do not have the knowledge, foresight, or incentives to effectively manage complex economic systems. The systems they plan will never be able to respond efficiently to the challenges that ultimately arise. And yet, even in the face of overwhelming evidence that their plans have failed, the political process forces them to place blame elsewhere and to use the power of their office to present themselves as the saviors of the situation. This may provide near-term relief to some, but it leaves the underlying problems in place and often exacerbates them.

The goal should be to minimize the influence that any government has on energy production or consumption choices. While there is a legitimate role for government in protecting public health and safety, that responsibility has been misused over time to justify broad interventions that do little to nothing to protect either of those things. That said, there is little evidence to suggest that American policymakers (or even many consumers)[33] recognize the benefits of broad energy economy liberalization, so it is imperative to find energy sources that can meet both the demands of a modern industrial economy and the political preferences of a majority of Americans.

AND WE FINALLY GET TO NUCLEAR ENERGY

This is where nuclear energy comes in. It is the one energy source that could meet the most demand preferences by the most people. It is abundant, reliable, and virtually emissions-free, and the United States has a long history of safe commercial nuclear operations. Indeed, almost 19 percent of the electricity Americans use comes from the country's ninety-four nuclear reactors—the most of any nation.[34] In terms of total production, the United States generates more nuclear energy than any other country. Globally, there are approximately 440 commercial nuclear power reactors currently in operation, with another 60 under construction and 90 more planned, most notably in China, India, and Russia.[35]

Nuclear power is clean. It produces gigawatts of electricity for millions of people while emitting virtually no air pollutants and requiring a small physical footprint. A single nuclear reactor like the ones used in the United States currently requires about $0.3 m_2$ of land per megawatt-hour, compared to wind ($8.4 m_2$ per MWh), solar ($19 m_2$ per MWh), and hydro ($14 m_2$ per MWh). This includes land used for mining, transportation, transmission, and storage. Nuclear is able to achieve this efficiency because of the density of its power.³⁶ A single pellet of uranium dioxide nuclear fuel about the size of a pencil eraser has as much energy as 150 gallons of oil, a ton of coal, or 17,000 cubic feet of natural gas.

Wind and solar energy enjoy a much better reputation as clean energy sources and also have benefits like zero emissions—assuming you don't consider the energy and resources needed to construct and maintain them. It should also be noted that there is a growing body of research showing that these energy sources are not as environmentally friendly as often portrayed. Regardless, they both require favorable weather conditions and backup power when the weather doesn't cooperate. If you need a power source to back up wind and solar, why build the wind and solar to begin with?

Nuclear reactors are online and generating power 93 percent of the time, compared with wind at 37 percent and solar at 26 percent.³⁷ And most nuclear power plants in the United States are licensed to operate for sixty years, while the operating life of intermittent energy like wind and solar is roughly half as long, and based on recent experience, it may even be less. Wind turbines will theoretically operate for twenty to twenty-five years, though some recent models suggest a lifespan of up to forty years. Performance degradation is also a concern, with one UK study finding that turbines can lose up to 13 percent of their normalized load factor within the first fifteen years of operation.³⁸ Even that is theoretical, given that the large-scale application of these technologies is still relatively new.

Like every energy resource, nuclear power does have its trade-offs. But even then, the reality is far better than we've been led to believe. Perhaps the first among people's concerns is safety. Many of us remain influenced by the infamous accidents at Chernobyl, Three Mile Island, and Fukushima. But should we be?

This may be hard to believe, but no one died or was even sickened by radiation exposure from the latter two incidents. In the case of America's worst nuclear accident, at Three Mile Island in 1979, the two million people living closest to the reactor received less radiation exposure than they would have from a dental X-ray. For decades, state and federal agencies and private companies tested agricultural, health, and environmental factors but found nothing of concern.³⁹

The accident at Chernobyl in 1986, which resulted from an egregious, unethical Soviet experiment, was more a commentary on authoritarian government than on nuclear technology. The Chernobyl reactor also lacked important safety features that are common to all US reactors, like containment domes. So far, the UN has confirmed forty-three deaths from radiation at Chernobyl, which is considered the worst nuclear accident in history.

Radiation itself is another common public concern, though one not broadly understood. Radiation is a part of our everyday lives. Flying in an airplane, eating bananas and carrots, sunbathing, getting medical scans, and simply living on earth all expose us to more radiation than living within fifty miles of a nuclear power plant. Radiation is an inherent and necessary part of life.

But fear has caused unnecessary environmental harm and costs. While visiting Fukushima, Michael Shellenberger, the founder of the pronuclear group Environmental Progress, challenged the Japanese government's colossal efforts to remove thousands of tons of "contaminated" topsoil to reduce radiation. The response he got was telling: "Every scientist and radiation expert in the world who comes here says the same thing. We know we don't need to reduce radiation levels … We're doing it because the people want us to."[40]

Another concern is nuclear waste. The US has over 90,000 metric tons of nuclear waste (in the form of spent fuel) from commercial power reactors, and that number increases by around 2,000 metric tons each year.[41] To put it another way, all the nuclear waste from every commercial reactor in the United States since 1957 amounts to no more than a football field ten yards high. For reference, the International Renewable Energy Agency estimates that by 2030, the United States will have between 170,000 and 1 million tons of waste from solar panels. The waste from renewable energy sources like wind, solar, and batteries can also be toxic and costly to manage.[42]

While the politics of nuclear waste management in the United States have delayed a solution, it is a technically solvable challenge. The nuclear industry in Finland, for instance, has demonstrated that one option is to build a deep geologic repository to permanently store nuclear waste.[43]

The point is not that nuclear power is perfect, but that it has a compelling track record despite public perceptions. All energy resources have trade-offs; there is no perfect resource. And nuclear power has some unique challenges. But it also has some incredible benefits that make it a choice well worth considering as a clean energy option.

IF CO_2 IS A PROBLEM, NUCLEAR IS THE SOLUTION

If we put aside global warming science and focus simply on the debatable idea that purposeful CO_2 reductions are good policy, the fact is that (absent some technological breakthrough) nuclear power is essential to meeting any real reduction goals. It is not just that nuclear power is needed but that a massive amount of nuclear power is needed in a relatively short period of time. Nuclear power provided 47.8 percent of America's carbon-free electricity in 2023.[44]

This is a critical point when we're discussing what's required to achieve substantial CO_2 reductions. Because wind and solar farms provide so little of America's energy, and the nation's electric system was built on reliable sources like coal, nuclear power, gas, and hydropower, policymakers could force some wind and solar into the system without it breaking down. But as reliable sources are replaced with intermittent ones, the system becomes less robust. If the objective is to push America toward massive CO_2 reductions while preserving the strength of its energy infrastructure, then nuclear energy must be the backbone of that system.

Let's agree that President Biden's net-zero policies are unattainable and instead take something more "realistic," like his 2030 objective to achieve a 50 to 52 percent reduction in carbon dioxide emissions from 2005 levels. To achieve that reduction, let's say we want half of America's power to come from carbon-free sources, while also maintaining the system's integrity. Assuming no growth in demand, which is admittedly ultra conservative, the US would have to *at least double* its current nuclear capacity to have any chance of meeting those goals without massive reductions in economic growth and standards of living.

1 GW/reactor

What would that look like? The US has ninety-four operating reactors today, with a total capacity of approximately 95 gigawatts (GW). Seven reactor designs are currently approved by the Nuclear Regulatory Commission (NRC). Most are large light-water reactors (LWRs); one small modular reactor (SMR) design is also approved. Though the prospect of SMRs has been exciting in recent years, the reality is that big energy demand will likely be solved by big nuclear power plants. Still, some combination of large, medium, and small reactors may be built.

Since reactors can be as large as well over a gigawatt of power and as small as under 50 megawatts (MW), we could assume an average of 750 MW per reactor. At that size, the US would need to build a minimum of 125 reactors by 2030. The truth is that number could be much higher because demand will grow and some of today's reactors will need to be replaced.

Could that be done? Sure. In fact, the United States did it before. America's first commercial power reactor came online in 1958 in Shippingport, Pennsylvania, having broken ground only about four years earlier. In its first year, it produced 60 MW of electricity.[45] By 1970, nuclear power provided around 6,334 MW of electricity, and by 1990, that amount had surpassed 96,000 MW, which is still about what is produced today.[46]

Some may argue that was then and this is now and no one could realistically build that many reactors in that short an amount of time today. Well, some never asked China about that. That country has added 34 GW of nuclear power over the past decade and has an estimated 23.7 GW of additional capacity currently under construction.[47]

But the reality is different in the US. Only two new commercial power reactors have been started and completed since the late 1970s—both at the Vogtle plant in Georgia—and the country does not have the industrial infrastructure to build even one reactor today.[48] The industrial and intellectual base atrophied as the nuclear industry declined over the past three decades. Large forging production, heavy manufacturing, specialized piping, mining, fuel services, uranium mining and enrichment, and skilled labor would all need to be reconstituted in massive quantities.

Global supply is no more promising, especially given that the rest of the world is coming to similar conclusions about the emerging role of nuclear power in meeting energy needs and CO_2 reduction targets. The global nuclear industrial base currently barely supports the sixty reactors under construction (mostly in Asia and Russia) and the normal operation and maintenance of the world's existing 440 reactors (including those in the US). Even global warming alarmists recognize that their carbon reduction goals are unattainable absent nuclear energy, and they have recently called for a tripling of global nuclear power capacity.[49] None of this means, however, that a massive nuclear build-out in the US is not possible. It just means that circumstances need to change to allow it to happen.

Without those changes, it's disingenuous to use such optimistic nuclear projections to argue that the entire power system can be overhauled with minimal economic consequences. Thus, any calls for a rapid increase in nuclear power necessarily must include radical policy reform that would allow the United States to engage in a commercial nuclear build-out like it undertook in past decades.

Many politicians and organizations attempt to remain agnostic toward nuclear energy by arguing that it might have a role to play if certain conditions are met. They then ensure that their conditions are set in such a way as to be unattainable. For example, they insist that the nuclear industry must improve its safety record, ignoring the fact that no one has ever been sickened or died because of commercial nuclear power in the US. How do you improve on this? Or they argue that the waste problem must first be solved but then stand in the way of safe methods of waste management, such as reprocessing or a geological repository.

If we view atmospheric emissions as such a threat that CO_2 reductions should be made the central organizing tenet of America's economic and energy policies, then the moral policy should be to achieve that objective in an economically rational way. The motives of anyone who would deny society access to the technologies most capable of achieving its stated goals, by either explicit antagonism or implicit passivity, must be questioned.

On the other hand, if CO_2 reduction is truly the objective, then maximizing America's nuclear resources as quickly as possible should be a top priority. While this would still not likely enable the US to meet the levels of nuclear power needed for Biden's CO_2 reduction goals, it could at least minimize the economic impact of his agenda. But doing so would require long-term, sustained bipartisan support for nuclear energy that results in major policy reforms. Without this, the billions of dollars of private capital needed to expand America's nuclear capacity will simply not be invested.

In reading the headlines today and accounting for the billions of taxpayer dollars currently being allotted to nuclear power, one could conclude that such support has been secured. But this would be a mistake. Support must not come in the form of subsidies, mandates, and government research. The United States tried this approach not long ago, and it failed.

In 2008, I wrote, in a report for the Heritage Foundation:

> Approximately 20 companies and consortia from around the world have recently released plans to build around 30 reactors in the United States. Some of these planned reactors may never be built. On the other hand, many more may be built. The U.S. is facing a 40 percent increase in electricity demand over the next 25 years. The pressure to reduce CO_2 emissions and dependence on foreign energy, combined with the inability of wind or solar power to meet the energy demand affordably or reliably, creates huge potential for nuclear power.[50]

This paragraph was describing the so-called, and hoped for, nuclear renaissance that many thought had been unleashed by a set of subsidies in the 2005 Energy Policy Act. Care to guess what those billions of dollars of public and private investment led to?

Exactly two new nuclear reactors.

CHAPTER 2
SECURITY THROUGH AFFORDABLE, CLEAN, SAFE ENERGY

Too often, Americans think of commercial nuclear energy dominance as a critical tenet of US power. But national pursuits disguised as commercial enterprises nearly always fail because political interests end up overtaking good economics. The result is a weak industry that consumes vital public and private resources. Luckily, nuclear energy can succeed on its own because of its unique characteristics. Unlike most any other energy source, nuclear combines affordability, cleanliness, and safety, and it does so primarily through domestic production.

AFFORDABILITY

No commercial enterprise will be successful over the long term if it doesn't provide a product or service that people want at prices they are willing to pay. The power of supply and demand will outlast political will every time. Small businesses and entire commercial empires have crumbled under this immovable truth, yet somehow, when it comes to energy policy, it is often ignored.

You don't need to look further than modern energy policy to see that. No amount of taxpayer dollars, subsidies, or mandates can create a sustainable market for electric vehicles, for example, so long as consumers prefer gas engines, and no amount of gaslighting can magically transform this reality to the government's preference.

This is also the case for electricity production. Governments have been trying to impose an energy transition away from hydrocarbons to renewables for decades. And despite the ongoing rhetoric from politicians, bureaucrats, and special interests that renewables are getting cheaper and more competitive all the time, anyone can see that this isn't the whole story. We know this because government continues to artificially inflate the cost of hydrocarbons while simultaneously subsidizing renewables to incentivize wide-scale uptake. As the flaws of this approach become apparent, politicians are turning more to nuclear power and applying the same subsidize-first mentality to that technology.

The problem, however, is that though nuclear energy has much to offer, it is still subject to the basic laws of economics. No matter how much some may want to expand America's use of nuclear, this will be virtually impossible if it's not economical on its own terms, and no amount of taxpayer support or subsidization will change that.

Many have concluded that nuclear is just too expensive to be the foundation for America's energy future. And who could blame them? Whether it's the new US reactors in Georgia or recent experiences in France and Finland, a mass expansion of nuclear energy does seem cost-prohibitive. But there are reasons to be optimistic.

The greatest of these is that nuclear energy has not always been so expensive. According to the Energy Information Administration, the overnight construction costs (costs as if they could occur overnight and not accrue interest) of a nuclear plant in 1967 were $600 to $900 per kilowatt (kW) in 2010 dollars. By the mid-1970s, the overnight cost range had exploded to $3,000 to $6,000. In 2023, the EIA estimated that a new plant in the US would cost $7,200 per kW (or about $5,800 in 2010 dollars).[1] EIA blames these increases on regulation, licensing problems, and poor project management, among other issues.

The costs of financing are another major expense for new nuclear plants—and the longer a project takes, the higher those costs climb. But reactors don't need to take a decade or longer to build. In the United States, they used to be built in about five years. Since time is literally money, the regulatory burden adds to the financing costs as well as the overnight costs of construction. Not only is the Nuclear Regulatory Commission too heavy-handed in its approach, especially for a technology that's been safely operating in the US for over half a century, but many of its regulations are based on antiquated ideas about radiation protection.

Unlike almost any other product, nuclear energy got more expensive as it became more abundant. This tells us that there's a lot of room for cost savings if we can get the right policies to allow what was done before to be done again. This is largely because most of the costs associated with nuclear power are in construction. Once built, nuclear plants are relatively cheap to operate because of the fuel they use, which is usually uranium. Nuclear fuel is incredibly energy dense. In other words, you don't need much, volumetrically speaking, to produce a lot of power.

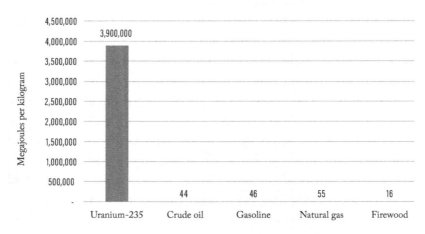

SOURCE: Visual Capitalist, 2023.

As the figure on the previous page shows, even when compared to relatively energy-dense hydrocarbons, nuclear energy is a clear winner, even though recent price increases for uranium ore and enrichment have reduced this advantage.

Another way to look at energy density is to consider the land requirements for different energy sources. This is especially helpful when comparing nuclear energy to intermittent sources like wind and solar, which are recognized as having low energy density.[2]

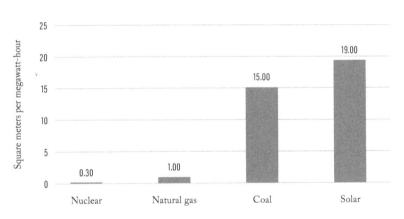

LAND USE BY ENERGY SOURCE

SOURCE: Our World in Data, 2022.

Again, nuclear is the clear winner.

Energy density alone, however, does not mean that an energy source is economical. There are any number of variables that contribute to the cost of producing energy. For example, if the fuel used is extraordinarily expensive, its density might not matter as much. On the other hand, if land is free and plentiful, then the amount of land used would not matter much.

As it turns out, the fuel for nuclear reactors is both relatively inexpensive and widely available. Most nuclear plants use uranium as fuel. Although the price of uranium has increased in recent years, the levelized costs of uranium-based fuel (or the total costs to produce energy per unit of output) are slightly less than the levelized costs of coal and substantially less than natural gas or oil.³ Though the all-in cost of an energy source doesn't tell the entire story of its value, it provides one useful point of comparison.

Electricity produced by nuclear energy is also minimally sensitive to uranium price swings because uranium accounts for only 15 to 20 percent of operating costs for nuclear power plants.⁴ According to the Nuclear Energy Agency of the Organisation for Economic Co-operation and Development (OECD), even a 50 percent change in nuclear fuel prices would only "slightly" impact the price of nuclear energy.⁵ On the other hand, combined cycle natural gas plants, which are the most efficient way to produce power using natural gas, experience a 7 percent shift in prices for every 10 percent change in fuel costs. Coal plants experience a 4 percent change in prices for every 10 percent rise in fuel costs. But it must also be acknowledged that nuclear fuel costs could become a larger issue if policies don't change to allow for a more seamless expansion of the uranium fuel supply chain to keep up with demand.

In a properly functioning market, with government staying out of the way, the uranium fuel industry would respond appropriately to market forces, even if fuel prices were to increase substantially. This means that price increases should drive industry to invest in developing additional uranium fuel supplies. And nuclear power doesn't end with uranium. Though the current industry is largely fueled that way, other reactor technologies use different fuel types, such as thorium, and some can even derive power from spent nuclear fuel.

The challenge facing America's domestic nuclear industry is that the country has lost its position as a global leader in the production of nuclear fuel, which has made it largely dependent on foreign suppliers. Accessing foreign supplies is not a problem per se. But it becomes a problem when foreign suppliers are not friendly, domestic alternatives do not exist, and onerous regulations make developing alternative fuel cycles all but impossible. This is exactly where the US finds itself today.

The problem is not that the US lacks the raw resources, the technological capability, or the industrial know-how to produce uranium or alternative nuclear fuel. It's that government interventions that make mining difficult, markets unpredictable, and construction of large industrial facilities costly have undermined the natural market incentives to expand the domestic industry to where it should or could be. Rather than fight Washington, most of America's nuclear fuel supply chain has moved overseas—an issue that will be explored in a later chapter.

Ultimately, though, the issue comes down to competitiveness, and that is generally a question not of fuel costs or operations costs but of construction costs. So can new nuclear plants be built and compete with other energy sources at unsubsidized price levels? To answer this, we need to identify what electricity sells for and what new nuclear costs.

The production cost of energy is only one component of the price consumers pay for their electricity. On average, energy generation represents about 60 percent of this price.[6] Transmission and other fees make up the other 40 percent. With an average price of approximately 13 cents per kilowatt-hour across all sectors (residential, commercial, and industrial), electricity, generally speaking, becomes profitable at about 8 cents per kilowatt-hour (or $80 per megawatt-hour [MWh]). If a power plant can produce electricity for less than that price, it will likely be competitive in a free energy market at current average prices.

Based on recent figures, new nuclear is expensive and questionably competitive. According to *POWER* magazine, the US reactors recently built in Georgia for a cost of $35 billion will produce electricity at $170 to $180 MWh.[7] In 1970, when nuclear technology was relatively new, a typical 1,000 MW plant cost $313 million to build (about $3 billion in 2024 dollars, adjusted for price inflation of construction materials).[8] At this price, assuming today's operating costs, nuclear power would have a levelized cost of electricity of $50.75 MWh.[9] This is far lower than the Vogtle plant and is competitive with other dispatchable baseload generation sources, such as coal and natural gas.

We should reject the idea that nuclear energy is necessarily expensive and instead draw from past experience to understand what prices *could* be. The next figure provides an estimate of the cost of electricity generated by nuclear power plants (if the cost of constructing these plants simply increased at the same rate as general construction costs) alongside the estimated cost of other technologies entering service in 2027.[10]

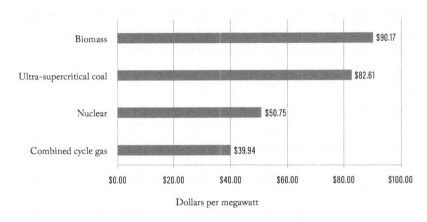

SOURCE: Energy Information Administration.

Some may argue that nuclear energy presents different safety challenges than other energy sources and is therefore going to be subject to additional regulatory burdens. While there could be some truth to that, the safety record of nuclear throughout its history, including prior to the ratcheting up of the regulatory burden, should provide confidence that we can develop an approach that is less onerous but equally safe.

ENVIRONMENTAL CONSIDERATIONS

Environmental controls present perhaps the greatest regulatory hurdle to future energy supplies. Ironically, the push for restrictions on CO_2 emissions is an opportunity for nuclear power.

Carbon dioxide emissions are the latest anxiety driving environmental activists. Their sophisticated public relations campaign has convinced much of the world that CO_2 and other naturally occurring gases, such as methane, cause catastrophic global warming and must therefore be drastically reduced. Both nationally and internationally, governments have imposed or are considering schemes to restrict the release of these gases. States are also responding to CO_2 fearmongering. California has even banned gasoline-powered lawn equipment, while other states are working on similar mandates.[11]

To avoid drastic consequences for the economy and for Americans' lifestyles, we must leverage an affordable energy source that can meet CO_2 objectives while also satisfying the needs of a modern industrial society. Because nuclear energy emits virtually no atmospheric pollutants, it is the best way to achieve these objectives.

Even President Biden acknowledges a role for nuclear power. In fact, in May 2024, Biden's White House made a major announcement about nuclear energy that included a laundry list of subsidies and programs meant to reignite America's commercial nuclear industry.[12] Unfortunately, the announcement was more about how government can subsidize nuclear power than it was about embracing reforms that would allow the industry to grow organically. But it did demonstrate the general recognition that nuclear is important.

One thing that Biden ignores almost completely, though, is the problem of managing and disposing of spent nuclear fuel, or nuclear waste. Solving this issue is critical because it's one of the main environmental sticking points that opponents of nuclear often bring up. But nuclear waste should actually be understood as an environmental selling point for nuclear energy because its by-products remain contained and can be harnessed for future use.

While environmentalists like to suggest that strict energy conservation mandates and a forced transition to intermittent energy sources like wind and solar are the answers to meeting future energy demands, these options are not as environmentally friendly as they are often portrayed. Even if they were affordable and capable of meeting growing demand (which is questionable), they would require huge swaths of land, massive mineral development, and a misallocation of financial, technological, and industrial resources. Each source also produces industrial waste, which is often overlooked.

And then there is land use. Using wind, solar, and biomass to meet future energy demand would devour rainforests, mountaintops, and shorelines. As mentioned earlier, wind power requires nearly thirty times more land than nuclear power, while installed on-ground solar uses about sixty-three times the land area to produce the same amount of electricity.[13] Brazil's reliance on biofuels is already leading to fears of deforestation of the Amazon and other biodiverse regions.[14]

By almost any measure, nuclear energy is about the most environmentally friendly energy source we have.

ADVANCING NATIONAL AND ENERGY SECURITY

Decreasing American dependence on foreign energy has been a critical piece of the debate for decades. But dependence on foreign sources of energy is not a problem per se. The problem is that the United States has created economic and strategic vulnerabilities by its overdependence on unstable foreign energy sources. These dependencies are the result of bad energy policy, and continuing down this path will exacerbate these problems.

According to the United States International Trade Commission, the country imported 78 percent of its rare earth materials from China in 2019.[15] These are the precise elements that are needed to produce almost all the energy technology government is imposing on us. Supply disruptions caused by geopolitical conflict could be disastrous. Almost all modern advanced technology, from everyday consumer electronics to solar panels and critical military systems, relies on rare earth elements. By increasing our dependence by virtue of EV mandates and other means of state compulsion, we are by definition increasing our reliance on China.

Of course, that could be mitigated by creating a business environment that allows for more domestic development, but to this point, there is clearly no appetite for that. And even if there were, absent some type of ban on Chinese imports, we would still increase our dependence at least in the near- and midterm. While a ban could lessen our dependence over time, limiting supply always increases prices. Given the already high costs of so-called green power sources, such a ban would exacerbate the underlying bad economics.

The fact is that we rely on hydrocarbons for nearly every aspect of our lives. That demand is not going away anytime soon, regardless of the left's rhetoric. So the push for wind and solar power, as well as electric vehicles—combined with efforts to limit, if not outright ban, domestic hydrocarbon production—will force the US to look elsewhere for its hydrocarbon supply "as a bridge" until green tech becomes more available. This is what the Europeans did, and we are seeing the results of that. Their dependence increased, which increased their vulnerability.[16]

Given current technology, expanding nuclear power in the United States is the only way to meet environmental objectives without creating society-wide economic harm and massive strategic vulnerabilities through energy dependence on adversarial nations like China and Russia.

But it is not all about protecting against dependence on other nations. There are also many foreign policy benefits to nuclear energy. As Russia has made clear and China is working toward, building relationships through cooperation on nuclear energy creates important long-term connections. The US is already doing this by expanding its exports of natural gas, and nuclear energy cooperation will amplify these benefits.

According to the World Nuclear Association, Russia currently operates nine reactors in four countries: China, Iran, India, and Belarus.[17] That's just the beginning, however. It is also constructing nineteen reactors in six countries: India, Bangladesh, Turkey, Iran, China, and Egypt. It doesn't end there. There are plans in place to build up to thirty more reactors in countries including Bulgaria, South Africa, Nigeria, Argentina, Indonesia, Algeria, Jordan, and Vietnam. Russia claims it can offer electricity generated from these reactors for no more than $50 to $60 MWh—a significant advantage over the US reactors recently built in Georgia.

But that's not Russia's only advantage over the United States. It's also able to offer the full spectrum of commercial nuclear services. Not only can it provide state-of-the-art reactors and financing (which the US can also do), but it can also offer nuclear fuel and waste management services (which the US cannot). In essence, Russia can offer a full-spectrum, turnkey nuclear energy package.

Although China's export business is not as mature as Russia's, it is working to emerge as a serious competitor and is honing its craft by building a lot of reactors at home. Of the sixty or so reactors currently under construction worldwide, China is building twenty-seven.[18] And while the United States built those two reactors in Georgia over the past decade, China built thirty-seven.[19] It should be obvious to anyone that China is preparing to be a major commercial nuclear exporter in the coming years.

Why should we care if China and Russia are supplying the world with commercial nuclear power while we debate how much public money can be spent to subsidize nuclear energy and what to do with nuclear waste? Because China and Russia understand that these relationships can last for decades and can be used to control foreign clients.

This might not matter for countries that are already under Russian and Chinese spheres of influence. But as demand for clean, abundant, and affordable electricity grows around the world, the demand for affordable nuclear energy will surely follow. If the United States and other like-minded countries have nothing to offer but overpriced and overbureaucratized options that don't even include credible fuel and waste management services, then more and more countries will turn to Russian and Chinese providers. They will do this not because they want to fall under the control of these nations but because they have no alternatives.

Luckily, Russia and China are not the only countries to have continued evolving their commercial nuclear industry over the past few decades. So too has one of America's closest allies—South Korea. Indeed, South Korea is one of the only market economies to figure out both how to build affordable reactors at home and how to compete with Russia internationally.

The United States and South Korea have a long-standing and deep commercial nuclear relationship. Indeed, many of the technologies that South Korea now depends on were developed in coordination with the US government and US industry. The American nuclear industry—and the economy more broadly—has already benefited from this relationship. For example, although the American bid to build commercial reactors in the United Arab Emirates ultimately lost to the South Korean contender, the American nuclear industry was heavily involved with the project because of the relationship between the two nations.

More important than the mutual benefits of the cooperative relationship up to this point, however, is how it evolves moving forward. The US nuclear industry has a lot to offer with its deep history in technological development, culture of entrepreneurship and safety, and world-leading operations expertise. South Korea also has much to offer with its growing experience in building new reactors, both at home and abroad, at more economical prices.

For the United States to have any chance to compete with Russia and China in the commercial nuclear space in the future, it must grow and nurture its relationship with South Korea. To optimize this relationship, however, the United States needs to treat South Korea as a commercial equal. It needs to recognize that South Korea has a mature and growing commercial nuclear industry, and that it has a legitimate need for fuel production and spent fuel management capabilities. This has been controversial in recent years, as some have argued that these technologies present a nuclear proliferation threat. While both enrichment and spent fuel processing can be used to produce materials for nuclear weapons, it just doesn't make sense that South Korea would somehow be building nuclear weapons under the guise of its commercial nuclear program. As nations have been showing for decades, they do not need a Western-sanctioned commercial program to build nuclear weapons.

Furthermore, South Korea's domestic commercial nuclear industry and its export programs are large enough to justify developing full fuel cycle technologies. It is also the case that not all technologies are the same. It's perfectly reasonable to develop a waste-processing technology that does not result in any weapons-usable material. Moreover, any program to process spent fuel or to enrich uranium would certainly be subject to rigorous international monitoring.

Rather than pose a proliferation threat, the commercialization of these technologies will provide important alternatives to Russian and Chinese commercial nuclear fuel services. While American companies will at times compete with South Korean ones, it's also easy to imagine situations where consortiums of US and South Korean companies would bid together to win future international competitions. The US has a similarly deep relationship with Japan's very successful and mature commercial nuclear sector, and that relationship must also be nurtured and expanded.

Russia and China see nuclear energy as a means to advance their national security. The United States must not only recognize this fact but do what's necessary to effectively counter it. To that end, policymakers must resist the temptation to take the same approach as either Russia or China, which is to say to attempt to develop a nuclear industry based on state sponsorship and government support—if not outright ownership of commercial enterprises.

The bottom line is that for Western democracies to meet the challenges posed by Russia's and China's aggressive commercial export policies, they must cooperate in ways that allow for the comparative advantages of each to contribute to comprehensive offerings in the global market. In the US, that means allowing for an innovative and competitive industry to be born out of market forces, nurturing collaborative relationships with friends and allies that are also leaders in commercial nuclear technology, and updating export policies to allow for the seamless and safe trade of commercial nuclear products and services.

CHAPTER 3
PLAGUED BY MYTHS

Cultural narratives are a primary tool used by special interests to advance their policy agendas. What we now term "fake news" has been around for a long time, and perhaps no industry has fallen victim to this dynamic more than commercial nuclear power. It has proven to be safe, clean, and affordable. It *can* be produced primarily from domestic sources and is scalable to provide however much power is demanded for almost any use imaginable. Despite this—or maybe because of it—a negative cultural narrative took hold decades ago.

For nuclear energy to move forward in the US, the myths that feed this narrative must be addressed simply and honestly.

MYTH #1: NUCLEAR ENERGY MAKES GLOBAL WARMING WORSE

In addressing this first myth, we will, with extreme hesitancy, take as true the mainstream global warming narrative. But even if that narrative is true, there is nothing to support the idea that nuclear power exacerbates or accelerates climate change.

Given that nuclear fission does not produce atmospheric emissions, those who perpetuate this myth are clearly on a carbon dioxide witch hunt that focuses on other emissions-producing activities surrounding nuclear power, such as uranium mining and plant construction. Finding fault with nuclear energy based on these indirect emissions (or co-benefits, as they are often described in regulatory parlance) is a tactic often used to make the industry seem more harmful than it is.

Whether activists like it or not, the nation runs on CO_2-producing hydrocarbons. Until the US changes its energy profile—which can be done with nuclear energy!—almost any activity, even building windmills, will result in CO_2 emissions. If coal-fired power plants had been used to produce the 775 terawatt-hours of electricity generated by US nuclear plants in 2023, they would have released 808 million metric tons of carbon dioxide into the atmosphere. Instead, thanks to emissions-free nuclear power, those tons of CO_2 were never emitted—the equivalent of taking 78 percent of cars off the roads.[1]

What makes nuclear energy so exciting from an environmental standpoint is not the pollution that it has prevented in the past, but its potential for enormous savings in the future. Ground transportation is a favorite target of environmentalists, and they are correct that America's transportation choices are a primary source of the nation's emissions. Electric cars, which require significant development to achieve subsidy-free market viability, are seen as a potential solution to the problem. But if the electricity comes from gas and coal power plants, the pollution is simply transferred from a mobile energy source to a fixed one. With a nuclear power plant, that's not an issue.

This is not to argue that the US should be replacing gas and coal power plants with nuclear ones as a matter of policy. It is simply to point out that efforts to stop nuclear based on an indirect-emissions argument make no sense.

MYTH #2: THERE IS NO SOLUTION TO THE PROBLEM OF NUCLEAR WASTE

This is perhaps the greatest myth about nuclear power, and it's used to keep plants from being built, to chase capital away, and to scare the public off supporting the industry. While it is true that the United States government has completely failed in developing a solution to nuclear waste, it is not true that no solutions exist.

In fact, there are multiple approaches to managing and disposing of nuclear waste, and new techniques would already have been developed if the federal government hadn't intervened. For example, spent nuclear fuel can be removed from a reactor, reprocessed to separate unused fuel and other valuable components, and then used again for nuclear fuel or for other commercial applications. The remaining waste could then be placed in either interim or long-term storage, such as in the permanent waste repository proposed at Yucca Mountain in Nevada. France and other countries carry out some version of this process safely every day. And technological advances could yield greater efficiencies and improve the process down the road. South Korea, for example, is working toward future commercialization of its own reprocessing technology, while American companies are developing reactors that can be fueled by waste and are even devising methods to extract valuable elements from spent fuel. The argument that there is no solution to the waste problem is simply wrong.

Recycling spent fuel—often referred to as closing the fuel cycle—would enable the US to move away from relying so heavily on the proposed geologic waste storage plan for the success of its commercial nuclear industry. This would allow a more reasonable mixed approach to nuclear waste, which would likely include some combination of permanent storage, interim storage, reprocessing, and new technologies. We will expand substantially on this subject in a later chapter.

MYTH #3: NUCLEAR POWER RELEASES DANGEROUS AMOUNTS OF RADIATION

This myth—one of the most pernicious—relies on taking facts completely out of context. By exploiting public fears of all things radioactive and not educating people about the true nature of radiation and radiation exposure, antinuclear activists can easily portray any radioactive emissions as a reason to stop nuclear power altogether. But when radiation is put into the proper context, the safety of nuclear power is clear.

Nuclear power plants do emit minimal amounts of radiation, but the amounts are environmentally insignificant and pose no threat. These emissions fall well below the legal safety limit sanctioned by the Nuclear Regulatory Commission (NRC) and the Environmental Protection Agency (EPA). The average annual radiation dose per person in the US—from everyday activities like getting an X-ray or being exposed to radon gas in the atmosphere—is 620 millirem (mrem). Roughly half of this is naturally occurring. Living near a nuclear power plant contributes less than 1 mrem.[2] Put another way, less than 0.2 percent of the average person's annual radiation dose comes from nuclear power plants.[3] Meanwhile, an airline passenger who flies from New York to Los Angeles receives 2.5 mrem on a single flight. As the chart below illustrates, radiation exposure is an unavoidable (and necessary) part of everyday life on earth, which itself emits low levels of radiation.

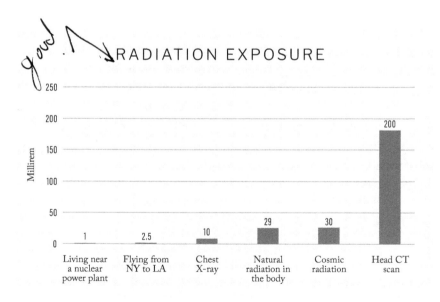

SOURCE: Environmental Protection Agency.

An even more pernicious result of this narrative about the dangers of radiation is that it's an obstacle to updating regulatory standards to reflect a modern understanding of the risks associated with nuclear energy. The existing regulatory approach relies on the precautionary principle and the linear no-threshold (LNT) model to assess those risks. The LNT model presumes that any dose of radiation, no matter how small, increases cancer risks, and therefore any exposure is dangerous.[4] When this is combined with the precautionary principle—which essentially says that in the absence of certainty that an action will or will not cause harm, any action should be avoided—the foundation for hamstringing nuclear energy has been laid.

These issues will be discussed more thoroughly in a later chapter, but for now, suffice it to say that <u>there is ample evidence that exposure to low doses of radiation does not present a significant threat to human health</u> and safety, and therefore LNT is an inappropriate approach to regulating radiation exposure.[5] And if that's the case, there is also no justification for applying the precautionary principle. That is not to say that radiation exposure should not be regulated, but the regulations should reflect the best scientific understanding of the risks posed.

MYTH #4: NUCLEAR REACTORS ARE VULNERABLE TO TERRORIST ATTACK

This is an especially salient claim in the post-9/11 world, where Americans have been conditioned to fear terrorist strikes on the homeland and are fed news stories of the vulnerability of nuclear power plants on a semi-regular basis. The truth, however, is that nuclear reactors are now designed to withstand the impact of airborne objects like passenger planes, and operators and regulators routinely review and update standards and procedures to protect against a myriad of threats.

A successful terrorist attack against a nuclear power plant could have severe consequences, as would attacks on schools, chemical plants, or ports. But fear of a terrorist attack is not a sufficient reason to deny society access to any of these critical assets.

Not one of the world's 440 commercial nuclear power plants has experienced a successful terrorist attack. As of this writing, even Ukraine's six-reactor nuclear power plant in Zaporizhzhia, which has been subject to Russian military operations and occupation, remains safe.[6] This site was even attacked by drones, yet safety was never compromised. Of course, history should not beget complacency, especially when the stakes are so high. But the Nuclear Regulatory Commission has worked with the industry to heighten security and increase safeguards on-site to deal with the threat of terrorism, and to be prepared should such an event unfold.[7]

The right response to terrorist threats to nuclear plants—like threats to anything else—is not to shut them down but to secure them, defend them, and manage the consequences in the unlikely event that an attack occurs. Allowing the fear of terrorism to obstruct the significant economic and societal gains from nuclear power is both irrational and unwise.

MYTH #5: NUCLEAR POWER LEADS TO NUCLEAR WEAPONS PROLIFERATION

This myth relies on creating an illusion of cause and effect. This is why so much antinuclear propaganda focuses on trying to equate nuclear weapons with civilian nuclear power. Once this spurious relationship is established, antinuclear activists can mix and match causes and effects without regard for the facts. Media outlets perpetuate the illusion by routinely conflating commercial nuclear power with the threat of nuclear weapons, which only serves to further misinform the public.

What's more, this argument is irrelevant inside the United States. As a matter of policy, the country already has enough nuclear weapons to hold any strategic target it desires at risk, so it's disingenuous to argue that expanding commercial nuclear activity in the United States would somehow lead to weapons proliferation. The same would hold true for any other country that has nuclear weapons.

It's also an oversimplification to suggest that granting countries access to peaceful nuclear technology will lead to weapons proliferation. Nuclear weapons require highly enriched uranium or plutonium, which can't be produced without a dedicated infrastructure. While most countries could certainly develop the capabilities needed to produce these materials, the vast majority clearly have no intention of doing so.

For start-up nuclear powers, whether the preferred method of acquiring weapons-grade material domestically is to enrich uranium or to separate plutonium from spent nuclear fuel, these processes are distinct from nuclear power production. Furthermore, countries that want to develop nuclear weapons capability will not be stopped, as demonstrated by North Korea and Iran. If proliferation is the concern, proper oversight is the answer—not stifling a distantly related industry.

It's a red herring to suggest that civilian nuclear programs provide cover for would-be nuclear states to build military nuclear capacity. Everyone already knows which countries want to build nuclear weapons, and these states are not hiding their intentions by pursuing alleged commercial capacity. No one was surprised that North Korea, Iran, or Pakistan built nuclear weapons. While these countries did pursue commercial programs (some certainly more than others) to build the expertise that may have contributed to their military nuclear weapons, it was common knowledge that they were doing that. The problem was not that they were using alleged commercial nuclear programs to hide their military intentions but that other nations lacked the political will to do anything about it.

MYTH #6: TRANSPORTING RADIOACTIVE MATERIALS EXPOSES PEOPLE TO UNACCEPTABLE RISK

A staggering amount of evidence directly refutes this myth. The truth is that the nuclear industry and the NRC and other regulatory agencies around the world take the strictest precautions when dealing with the transportation of nuclear materials, which the record demonstrates. Nuclear waste, for example, has been transported on roads and railways worldwide for years without a significant incident. A 2016 article on this subject concluded, "In the US alone, over one million miles of SNF [spent nuclear fuel] shipments have been safely conducted. Worldwide, the number of SNF shipments that have been made is not significantly less than 20,000, but due to incomplete records, may be much greater. Similarly, at least 35,000 tons of SNF have been shipped worldwide over the last 50 years or more, but the total quantity shipped may be greater."[8] Transportation of radioactive materials is just not a problem.

The NRC has six key components for safeguarding nuclear materials in transit:[9]

[handwritten margin note: Military like convoys]

1. Use of NRC-certified, structurally rugged overpacks and canisters
2. Advance planning and coordination with local law enforcement along approved routes
3. Protection of information about schedules
4. Regular communication between transports and control centers
5. Armed escorts within heavily populated areas
6. Vehicle immobility measures to prevent movement of a hijacked shipment before response forces arrive

Of course, the argument can always be made that even though there's a low likelihood of an accident happening, the damage caused would be so great that the risk is simply not worth it. This is a fatalistic argument that could be used to stop almost any progress on anything, and what's more, it's not supported by the facts.

There are two important points to keep in mind. First, spent nuclear fuel is not the sludgy, fluorescent green liquid often portrayed in the media. According to the Department of Energy, "The fuel used in today's commercial reactors is made up of small ceramic pellets of low-enriched uranium oxide. The fuel pellets are stacked vertically and encased in a metallic cladding to form a fuel rod. The fuel is a solid when it goes into the reactor and a solid when it comes out."[10] Once removed from the reactors and cooled in pools, the spent fuel is housed in steel and concrete containers that are extraordinarily robust. While an accident could certainly require cleanup, it's very unlikely to be the environmental disaster that's often portrayed.

Second, and something not often talked about, the commercial nuclear accidents that we have had were much less severe than originally feared. Consider the accident at the Fukushima Daiichi power plant in Japan. It's hard to imagine a worse accident than that, yet according to the World Health Organization, the health risks associated with the release of radioactive contaminants are low in Japan and extremely low in neighboring countries.[11] That's not to minimize the real economic and mental health impacts of the accident, but it is important perspective to bear in mind when considering the actual public health risks associated with nuclear power broadly and the transportation of nuclear materials specifically.

MYTH #7: NUCLEAR ENERGY IS NOT ECONOMICALLY VIABLE

This claim, while more complicated, is nonetheless a myth. Nuclear energy already provides about 19 percent of America's electricity and enjoys ongoing private-sector investment. But it is also true that some power plant operators claim they require subsidies to stay open, and that many nuclear advocates claim subsidies are necessary to bring new plants and technologies to the market.

Investors are not averse to nuclear power. Utility companies with nuclear experience have sought to purchase existing plants, upgrade their own plants, and extend their operating licenses so that they can produce more energy for a longer time. Indeed, nuclear energy is so economically viable that it already provides a substantial percentage of America's electricity despite the incredibly high regulatory burden, as noted earlier. This is not even to mention the substantial investment that's flowing into nuclear energy technology from entrepreneurs seeking to develop advanced reactors.

But investors are understandably averse to the regulatory risk associated with building new plants. The burden is extreme and potentially unpredictable. In the past, opponents of nuclear power have successfully used the regulatory process to file legal challenges simply because they oppose nuclear power. These challenges have caused considerable delays, resulting in increased construction costs.

So the subsidies in legislation like the Inflation Reduction Act are arguably necessary not because the market has rejected nuclear power but because it has rejected the excessive regulatory risk and costs imposed by the government and antinuclear activists. When making financing decisions, investors must consider the massive costs and losses caused by past government intervention. Until new plants have been constructed and are in operation (thereby proving that regulatory obstacles have been mitigated, both financially and legally), the burden of proof will remain on government regulators.

But the bigger picture on nuclear economics is that no one truly knows how economical (or not) nuclear power could be if it were allowed to exist in a free market. To achieve free-market conditions for nuclear energy, the federal government must create a predictable and reasonable regulatory environment, enact a market-based waste-disposition program, and stop subsidizing specific technologies and firms. Perhaps most importantly, Washington needs to stop subsidizing nuclear energy's competitors. Such an environment would allow capital to flow to the most promising technologies and allow the emergence of business models that promote safety, efficiency, and profitability over politics. Only then will we know the true economics of nuclear energy.

But beyond all that, we should always remain skeptical whenever a politician or bureaucrat advocates for specific energy policies based on cost. The role of policy should be to protect public health and safety and to set the rules and regulations that allow all energy sources to compete. Supply and demand driven by innovation, competition, and consumer preference within a free market is the only sure way to determine cost.

MYTH #8: NUCLEAR POWER IS INHERENTLY UNSAFE

This is perhaps the greatest myth surrounding nuclear power. But the reality is that past incidents in the US have been minor and their consequences limited, demonstrating that nuclear power is safe. This final myth depends on flawed logic and misrepresentations that are riddled with factual errors. The narratives generally unfold as follows:

- First, the consequences of Chernobyl are overblown to invoke general fear of nuclear power.
- Next, the Three Mile Island accident is falsely equated with Chernobyl to create the illusion of danger at home.
- Finally, any accident, no matter how minor, is portrayed as being ever so close to catastrophe so as to demonstrate the dangers of new nuclear power.

This myth can be dispelled simply by revisiting the real consequences—that is, the actual harms to public health and safety—of Chernobyl, Fukushima, and Three Mile Island. Although any loss of life is a tragedy, a more realistic presentation of the facts would show that these accidents demonstrate the inherent safety of nuclear power. Chernobyl was the result of government corruption, human error, and poor design. Of the fewer than fifty fatalities, most were rescue workers who unknowingly entered contaminated areas without being informed of the danger. In 2005, the World Health Organization said that up to four thousand fatalities could ultimately result from Chernobyl-related cancers, but so far, this has not happened.[12] The primary health effect was a spike in thyroid cancer among children, with roughly five thousand diagnosed between 1992 and 2002. Of these, fifteen children tragically died, but 99 percent of cases were resolved favorably. No clear evidence indicates any increase in other cancers from exposure to radiation among the most heavily affected populations. Of course, this does not mean that cancers could not increase at some future date, nor does it detract from the human tragedy caused by the Soviet Union's mishandling of the Chernobyl accident. But today, almost forty years on, the expected long-term radiation deaths have not materialized.

Interestingly, the World Health Organization has also identified a condition called "paralyzing fatalism," which is caused by "persistent myths and misperceptions about the threat of radiation."[13] In other words, the propagation of ignorance by antinuclear activists has caused more harm to the affected populations than the radioactive fallout from the actual accident.

The most serious accident in US history involved the partial meltdown of a reactor core at Three Mile Island, in Pennsylvania, but no deaths or injuries resulted. The local population of two million people received an average estimated radiation dose of about 1 mrem—insignificant compared to what we all receive through our daily activities each year.[14]

The most recent accident was the one at the Fukushima Daiichi power plant in Japan. That accident occurred on March 11, 2011, when a tsunami triggered by a major earthquake damaged the plant's backup generators, causing the cooling systems in the operational reactors to fail. This led to the overheating and partial meltdown of the fuel rods, which in turn resulted in the release of radiation and the evacuation of the surrounding area.[15] Again, there was real economic harm and a psychological toll on local populations, but the impact of radiation exposure on physical human health was almost nonexistent.

Other smaller incidents have also occurred, and all have been resolved safely. For example, safety inspections in 2002 revealed a hole forming in a vessel-head at the Davis-Besse plant in Ohio.[16] Although only an inch of steel cladding prevented the hole from opening, the NRC found that the plant could have continued to operate for another thirteen months, and the steel cladding could have withstood pressures 125 percent above normal operations. Correctly, the plant was shut down until the problem could be fixed.

The collapse of a partial cooling tower at the Vermont Yankee plant was far less serious than the Davis-Besse incident but is nonetheless presented by activists as evidence of the risks posed by power reactors. Non-radioactive water was spilled in the collapse, but no radiation was released.[17]

As for vulnerability to earthquakes, the NRC requires each nuclear plant to meet a set of criteria to protect against earthquakes. The efficacy of these criteria was demonstrated in 2011, when two reactors at the North Anna power station in Virginia were subject to an earthquake and aftershock that measured 5.8 and 4.5 on the Richter scale, respectively.[18]

Even though the quake exceeded a number of bases for which the plant was designed, the safety systems were not damaged, and all equipment and personnel responded without issue. Though the power plant had to be shut down following the quake, it reopened about three months later.

In each of these cases, antinuclear activists portrayed these incidents as proof that nuclear energy is dangerous. But the facts reveal that the opposite is true. When incidents arise, they are identified and safely resolved. That is not an indication that a technology is dangerous but a confirmation that an industry is structured to assess risk and identify and effectively resolve anomalies.

Antinuclear activists successfully stopped the industry once before, but nuclear energy is too important to America to allow that to happen again. Despite opponents' efforts to mislead the public, nuclear energy has proven to be a viable, economical, and environmentally sound solution to growing US energy demand.

CHAPTER 4

THREE MILE ISLAND, CHERNOBYL, AND FUKUSHIMA DAIICHI

Although the regulatory and policy obstacles to nuclear energy are significant, they don't exist in a vacuum. Despite growing public ease with the technology, there remains an underlying cultural anxiety about nuclear power that is driven primarily by our collective memories of the accidents at Three Mile Island, Chernobyl, and Fukushima. But what really happened in these cases?

Let's take a look at each incident to see how it unfolded and what lessons were learned.

THREE MILE ISLAND

At about 4 a.m. on March 28, 1979, a cooling circuit pump in the non-nuclear section of Three Mile Island's second station (called TMI-2) malfunctioned, causing the reactor's primary coolant to heat and its internal pressure to rise. Within seconds, an automated response system had shut down the core. An escape valve opened to release pressure, but then it failed to close properly. A series of human and mechanical errors followed.

Control room operators—who could see that a "close" command had been sent to the relief valve—had no way of knowing the valve's actual position. With the valve open, coolant escaped through the pressurizer, and the operators received a signal that there was too much pressure in the coolant system. They shut down the water pumps to relieve that supposed pressure, which allowed coolant levels inside the reactor to fall (referred to in the industry as a loss-of-coolant accident). As the water dropped, the uranium core was exposed and became intensely hot, and eventually it began to melt.

By the time the operators discovered what was happening, superheated and partially radioactive steam had already built up in auxiliary tanks. The operators moved this steam to waste tanks through a series of compressors and pipes. But the compressors leaked, allowing hydrogen and other radioactive gases to escape the core. The leak was minuscule—an average estimated dose of about 1 millirem—and once discovered and addressed, it posed no ongoing threat to human health. This is not to minimize the seriousness of any unplanned radiological release but to put it into real-world context.

In the immediate aftermath of the accident, sample testing of air, water, milk, vegetation, and soil showed that there were negligible effects, and experts concluded that the radiation was safely contained. Numerous studies have since investigated the incident's health impact and have found no adverse health effects or links to cancer.[1] But Three Mile Island did have an enormous impact on American public opinion, and that led directly to onerous new regulations and contributed to a complete halt in the building of new plants. The NRC didn't issue another reactor construction permit until 2012.[2]

More positively, the nuclear industry and the NRC implemented a number of technological and procedural changes to considerably reduce the risk of a future meltdown. These included upgrades to design specifications and equipment requirements to improve things like the auxiliary feedwater systems and automatic shutdown capabilities. Operator training and staffing requirements were enhanced, and fitness-of-duty programs (to guard against substance abuse) were introduced. Transparency was improved through the release of periodic public reports and closer coordination among federal, state, and local agencies. The NRC's resident inspector program was expanded to require that at least two inspectors live near and work at each nuclear power plant in the US, enabling them to ensure daily adherence to NRC regulations. And the industry itself created the Institute of Nuclear Power Operations, a nonprofit organization that evaluates plants, promotes training and information sharing, and helps individual power stations overcome technical issues.[3]

CHERNOBYL

Seven years after the incident at Three Mile Island, on April 25, 1986, a crew of engineers with little background in reactor physics began an experiment at the Chernobyl nuclear station. They wanted to determine if the momentum of the plant's turbines could provide power if the main electrical supply to the station was cut. To carry out the experiment, operators chose to deactivate the plant's automatic shutdown mechanisms.

The four Chernobyl reactors were known to become unstable at low power settings, which is exactly what happened during the engineers' experiment. When the operators cut power and switched to the energy from the still-rotating turbines, the coolant pump system failed, causing heat and extreme steam pressure to build inside the reactor core. The reactor experienced a power surge and exploded, blowing the lid off the building. (The Chernobyl reactors lacked fully enclosed containment buildings, a basic safety installation for commercial reactors in the US.) For the next nine days, radioactive gases and flames spewed into the atmosphere, and this radioactive material was carried by the wind to most of the rest of Europe.

Of the fifty or so direct fatalities, most were rescue workers who'd entered contaminated areas without being informed of the danger.[4] In 2005, the World Health Organization and the International Atomic Energy Agency estimated that up to four thousand people could ultimately die from Chernobyl-related cancers.[5] These deaths could still emerge, but they have not materialized yet. The primary documented health effect was a spike in thyroid cancer among children, with as many as five thousand diagnosed between 1992 and 2002. Of these, fifteen children died.[6] There is no clear evidence of any increase in other cancers among the most heavily affected populations.[7]

As noted, the World Health Organization has also identified what it characterized as "widespread psychological reactions to the accident, which were due to fear of the radiation, not to the actual radiation doses."[8] This condition has been referred to as "paralyzing fatalism," and it must be part of any discussion of the long-term health effects of Chernobyl. That is certainly not to argue that people's lives weren't impacted in severe and tangible ways. They clearly were. But from a pure mental health standpoint, it seems that the culture of doom and dread that remains pervasive in the region has more to do with fears about the dangers of the long-term impact of Chernobyl than any actual threats to public health. It is worth imagining what the region might be like today if the cultural narrative around nuclear energy had historically been positive and our collective understanding of the risks posed by low-dose radiation was more realistic. It's impossible to know for sure, but you could reasonably conclude that the region would be much better off.

The Chernobyl accident was the result of both human error and poor design. It's not realistic, though, to compare the technology to that used in US commercial power reactors. First, the graphite-moderated, water-cooled reactor at Chernobyl generally maintained a high-positive void coefficient, which means that under certain conditions, the reactor's power output (or reactivity) could rapidly increase as its coolant heated (or was lost), resulting in more fissions, higher temperatures, and ultimately meltdown. This is in direct contrast to the light-water reactors used in the US, which would shut down under such conditions. American nuclear plants use water to both cool and moderate the reactors. The coolant keeps the temperature from rising too much, and the moderator is used to sustain the nuclear reaction. As the reaction occurs, the water heats up and becomes a less efficient moderator (cool water facilitates fission better than hot water in light-water reactors), causing the reaction to slow and the reactor to cool. This makes light-water reactors inherently safe, and it's one reason why a Chernobyl-like reactor could never be licensed in the US for commercial power production purposes.

After Chernobyl, many technological changes and safety regulations were put in place to prevent a similar meltdown from occurring. Reactors that used comparable technology in Russia, Ukraine, and Lithuania were renovated to make them more stable at lower power, increase the responsiveness of automatic shutdown operations, and automate and upgrade safety mechanisms. Chernobyl also led to a number of international efforts to promote plant safety through better training, management, and coordination. The World Association of Nuclear Operators, a nonprofit whose members operate over 450 nuclear reactors, is just one organization that grew out of these efforts.

FUKUSHIMA DAIICHI

The world had not seen a major accident at a commercial nuclear facility in twenty-five years when a devastating tsunami hit Japan in March 2011, killing over eighteen thousand people.[9] The human toll of the disaster was terrifying, as was what happened at the Fukushima Daiichi nuclear plant in its aftermath.

The incident began when an earthquake hit Japan on March 11.[10] The three operating reactors responded as intended, with all automatically initiating the shutdown process (the plant's other three reactors were closed for scheduled maintenance). Although the plant lost its connection to the grid, its backup generators provided power to continue a safe shutdown in the minutes following the quake. The problems started when the plant was hit by a large tsunami triggered by the quake. The waves were so high that they easily surged over the power station's protective seawall and effectively flooded the entire plant, damaging the emergency generators on the lower levels. One surviving backup generator still provided enough power to allow two of the three reactors to continue safely shutting down. Ultimately, however, nearly all the power needed to cool the reactors and allow for safe shutdowns had been lost, and all three reactors suffered at least partial meltdowns. This resulted in a buildup of heat and pressure, which in turn led to leaks of radioactive gas and hydrogen. It was the buildup of the leaked hydrogen that led to the iconic images, seen on

TVs around the world, of explosions at the three reactors. A̲ explosions did cause the release of radioactive material, it's im̲ note that they were hydrogen explosions that were the result of a rea̲ between the zirconium cladding of the fuel and the water, not nuclea̲ explosions in any way, shape, or form.

The reactors were stabilized within two weeks of the accident and were being safely cooled within four months. To date, Japanese authorities continue to work toward the safe decommissioning of the reactor sites and environmental cleanup.

As a result of the accident, the NRC mandated that US operators implement a series of upgrades at plants that use similarly designed reactors. These included things like installing additional emergency equipment, upgrading monitoring equipment, and adding venting systems to relieve pressure that could build during a Fukushima-style incident.

The story of what happened at the Fukushima Daiichi plant is complex. It would be easy to blame an act of God for the disaster, but it's not that simple. The danger of building any nuclear facility in a seismically active area is well understood, and the plant was designed to withstand earthquakes and tsunamis. But the design basis on which the plant relied was more than fifty years old, and the protection measures that existed at that time were inadequate. Despite this—and even though people had developed a better understanding of the site's vulnerability—neither the government nor the private sector acted to update Fukushima's tsunami protection measures beyond the original requirements. This resulted in several failures, but the most important was that the backup generators that would have allowed for the safe shutdown of the reactors remained on low floors, making them susceptible to flooding. Also, the tsunami protection walls were never heightened. Notably, other reactors in the region that had higher seawalls survived the tsunami without major incident.

The technical and operational lessons from the accident are critical, but we must not forget that the actual public health and environmental damage was limited. As terrible as the images of roofs flying off the reactors were, the fact is that no member of the public died because of radiation exposure due to the accident. Indeed, by the end of May 2011, no harmful health effects had been found in over 195,000 residents living

ted that one worker died in 2018 of lung
adiation exposure.¹² The vast majority of
radiation doses within normal limits or
ling to the World Health Organization, the
y the accident was not radiation exposure at
ess and the perception of radiation risk."¹³ In
alth risks associated with both Fukushima and
atest commercial nuclear accidents—have little
to do with u..,   iealth impact and everything to do with our own
perceptions of what happened.

HOW POLICYMAKERS RESPOND TO NUCLEAR ACCIDENTS

While hindsight is twenty-twenty, the evidence does suggest that had those responsible for the Fukushima plant acted in accordance with the evolving understanding of its vulnerabilities, the severest consequences of the tsunami could have been avoided. The incident demonstrates the importance of ensuring that reactor protection protocols always remain up to date and consistent with the latest science. In some cases, this could result in additional regulatory mandates, but in others, this could mean regulatory relief.

One of the problems with the dialogue that emerged in the US following the accident at Fukushima was that some policymakers assumed (or perpetuated the false narrative) that America's nuclear industry and regulatory bodies and policies mirrored Japan's. They did not. The United States had an effective, multifaceted regulatory regime that had already addressed many of the mistakes and weaknesses that Fukushima exposed, including earthquake and tsunami preparedness and the need to modify older reactors to meet new and evolving safety standards.

While building nuclear plants to withstand earthquakes and tsunamis (and other severe natural phenomena) was a new issue for many Americans to contemplate in the wake of Fukushima, the US nuclear industry and its regulators had spent a great deal of time developing specific protocols for just such events. Yes, there are US reactors in earthquake zones, but they were built to withstand not only the most powerful earthquake ever recorded for their respective sites, but also the strongest that geologists think are *possible* for each site.[14]

Moving forward, government regulators and policymakers must ensure that additional regulations promote true safety, not just the perception of safety. They must recognize that plant owners and operators are highly motivated to maintain safe operations and are in many ways better prepared to ensure public health and safety than federal regulators. Indeed, industry demonstrated its commitment to safety following Fukushima, just as it did following Three Mile Island: by reviewing its internal protocols, practices, and procedures and developing a response that promoted careful operations and made business sense.

It did this by creating a series of diverse and flexible coping strategies, known as FLEX.[15] FLEX provides guidance on how nuclear power plants can protect public health and safety even when threatened by events that exceed what a plant was designed to withstand. This approach is why the American nuclear industry has maintained a stellar safety record for the entirety of its existence. The industry consists of a competent, skilled workforce that is properly incentivized to maintain safe operations even beyond regulatory demands. This is not by accident.

Under current US policy, the plant operators are primarily responsible for safety. This means that if something goes wrong, the person in charge of that plant will be held accountable. Furthermore, the people who work in the plants also live in the local communities. This creates a powerful incentive to ensure that each plant is operating under the safest conditions possible. That's why the best approach has always been for nuclear regulators to set and enforce high standards based on actual risk assessments and allow plant operators to determine how best to meet them.

Unfortunately, in the months following the Fukushima accident, regulators and antinuclear activists around the world pounced. Rather than work to understand what happened and respond dispassionately, they used the incident as an excuse to justify more regulation—and even end the use of commercial nuclear power entirely. That was a mistake.

It is important to remember this, even more than a decade later, for a couple of reasons. First, some jurisdictions used the Fukushima accident to implement policies to pacify decades-old antinuclear sentiment. The best example of this is Germany, which decided to phase out nuclear power following Fukushima. California is also a great example. It used the Fukushima accident to build a case to shut down the Diablo Canyon nuclear plant, which is within a mile of the Shoreline fault.

Interestingly, the cost of these unwise decisions was quickly felt. In Germany, the consequences of shutting down major portions of its reliable electricity infrastructure while increasing its dependence on unreliable renewables and Russian natural gas became quite clear when Russia invaded Ukraine. As a result, Germany has since flirted with the idea of reopening some of its nuclear plants. California also became acutely aware of the flaws of its decision to shutter Diablo Canyon when it faced the threat of brownouts and blackouts and quickly reversed course. Diablo Canyon will remain open until at least 2030.

While nuclear energy is currently enjoying broad and growing support, that sentiment could shift quickly if another accident were to occur. The consequences of rash decision-making in the wake of such accidents could have a decades-long impact. Having said that, any negligence should be appropriately prosecuted, and authorities should seek lessons that can be applied more broadly to increase safety. But each incident must be dealt with discretely, and the temptation to draw broad conclusions about an entire industry or technology must be avoided.

ACCIDENTS DO NOT NECESSARILY REVEAL SYSTEMIC ISSUES

Policymakers should not assume that accidents such as those described above necessarily reflect deficiencies in the US regulatory system, which consists of both public and private regulatory entities. While there is plenty of room for criticism of the regulatory morass that any new nuclear project must navigate, the approach to maintaining safe operations at existing plants in the United States, as a general proposition, provides a sound foundation moving forward. Public and private regulatory bodies work in tandem to provide constant oversight, assessment, and evaluation of commercial nuclear operations.

Importantly, there is no institutional bias that keeps any of these structures—or the people within them—from raising questions or demanding action. "See something, say something" is not just a slogan for the American commercial nuclear industry—it is foundational to its culture. That's why such vigilance is encouraged. This doesn't mean that nothing ever goes wrong. But it does mean that anomalies are identified and mitigated in a timely manner.

US nuclear plant operations are also heavily influenced by the culture of safety instilled by the Naval Nuclear Propulsion Program. Many nuclear plant workers come from the navy and have absorbed the principles of this program, which is responsible for designing and building nuclear-powered ships and supporting the nuclear-powered naval fleet. While the American regulatory system is far from perfect, the results from a safety standpoint are close to it.

The way it works is that federal regulators at the NRC license commercial nuclear facilities and operators, develop protocols, and provide oversight and enforcement of those protocols. This process is supplemented by private self-regulation overseen by the Institute of Nuclear Power Operations (INPO). In addition to its work evaluating plant operations and training employees, INPO also provides technical and management assistance to owners and operators. And the operators themselves promote safety on specific sites, as well as in the industry more broadly. This combination of government and private regulation creates a complementary system that encourages safety from the macro level all the way down to the individual plant employee. For example, plants have employee training programs in place, and INPO regularly audits those programs to recommend improvements. This process helps ensure that federal safety guidelines are met and plants are operated extremely safely and efficiently.

The Nuclear Regulatory Commission is responsible for setting and enforcing safety standards, but plant owners have primary responsibility for operations. Ultimately, they benefit financially from safe workplaces.

More importantly, a system in which private owners are responsible for safe operations with strict government and private oversight creates a dynamic regulatory environment that encourages questioning and demands responses. Although accidents do happen and safety lapses do occur, the American system is designed to identify irregularities and resolve them quickly. As lessons are learned, plant operators, private institutions, and the NRC all work together to address shortcomings.

And the proof is in the pudding. The NRC keeps all sorts of data on safety trends at nuclear power plants, and the trends are unmistakable. An already safe industry is getting safer with each year that passes.[16]

Still, the American system can be overly bureaucratic and suffers from federal micromanagement. Many industry-driven safety advances, such as those engineered into new reactor designs, must traverse the federal bureaucracy before they can be brought to market. This is just one more barrier that makes new plant construction difficult, and it's something that should be addressed as a matter of general nuclear policy reform.

CHAPTER 5
THE BROKEN POWER PLANT FALLACY

Cost overruns in Europe and the United States provide plenty of fodder for critics who argue that nuclear energy is just too expensive. Take Finland's Olkiluoto 3 reactor as an example. It eventually came online in 2023, but it was $8 billion over budget and fourteen years behind schedule.[1] France's Flamanville 3 reactor has suffered a similar fate: it cost at least four times original estimates and ran the better part of a decade behind schedule.[2] The Vogtle AP1000 reactors in Georgia also took years longer and cost billions more than expected. And yet, delays and cost overruns are not necessarily indicative of the future economic viability of nuclear power.

There is a lot of blame to go around for the cost overruns and time delays for these projects, but it's important to note that each was essentially a first-of-a-kind project. It wouldn't be accurate to assume that future similar plants would cost as much. Construction costs should also be reduced as lessons learned from initial projects are integrated into future ones. Indeed, some estimates indicate that the second new reactor built at the Vogtle site was 30 percent cheaper than the first one.[3]

Some of the overruns are simply a reflection of rising labor and material costs. These increases, which are not unique to the nuclear industry, would affect any large-scale project today. A lack of skilled personnel, shortages of nuclear-qualified components and materials, and inexperienced vendors and subcontractors also slowed progress. Very few reactors have been built in Western countries over the past three decades, and the industrial base and skill sets needed to support a growing demand for commercial nuclear power have yet to mature. These risks should have been expected for these initial projects, but they're also correctable and will be resolved by the market over time (if allowed to do so).[4]

THE US EXPERIENCE

The construction of units 3 and 4 at Vogtle began in 2009, with the reactors originally expected to cost $14 billion and begin commercial operation in 2016 and 2017, respectively. But the project faced significant delays, and the total costs came in at over $30 billion.

Technically, the Vogtle reactors are the first to come online since 2016, when Tennessee's Watts Bar 2 reactor entered service, but this is somewhat misleading. Work on Watts Bar 2 actually began in 1973, was halted in the mid-1980s, and then was restarted in 2007.[5] Most of the reactors in the United States today, aside from the Vogtle reactors, were built between 1967 and 1990, and nearly all of them began construction in the 1960s and 1970s. The Vogtle reactors were the first new US nuclear plants built in about three decades.

Despite the challenges of restarting the American nuclear industry, bipartisan support emerged in the early 2000s to help kick-start the process. Ultimately, a suite of subsidies for commercial nuclear power were included in the wide-ranging Energy Policy Act (EPACT) of 2005, which also included a bonanza of other energy subsidies.[6] When it came to nuclear—and more specifically, spending meant to promote the reemergence of the industry—the subsidies included loan guarantees, risk loss insurance, cost-sharing programs, and production tax credits.

Largely as a result of the legislation, eighteen applications for combined licenses (COLs) to construct and operate twenty-nine new reactors were submitted to the Nuclear Regulatory Commission.[7] Unfortunately, despite the initial enthusiasm, political support, and taxpayer funding, most of the COL applications were eventually set aside, withdrawn, or terminated. But not all of them. Eight licenses were ultimately issued for the construction of fourteen reactors, but only the two reactors at Vogtle went on to be built.[8]

In 2010, the Obama administration announced $8.3 billion in loan guarantees—later increased to a total of $12 billion—to help artificially lower borrowing costs for project financiers. Let's be clear that federal loan guarantees do not actually lower costs—they simply socialize the costs among American taxpayers. In 2009, Georgia also passed a law allowing Georgia Power to charge customers for financing costs during the construction phase.[9]

Given this track record, it's difficult to argue that the subsidies helped the industry over the long term. On the surface, it looks like they at least helped one utility build two reactors. But even that is not clear. Enthusiasm around nuclear energy had already ignited private interest in constructing new plants prior to EPACT 2005. What might have happened if, instead of spending time, money, and effort on figuring out how to maximize access to taxpayer support, public and private nuclear energy advocates had focused on developing innovative financing options, solving supply chain challenges, and resolving regulatory issues that continued to slow progress?

Of course, no one can answer that question, but what we can say is that the EPACT 2005 subsidies did not usher in a nuclear renaissance. What's worse, they further calcified the dependency relationship that the nuclear industry, by and large, has with Washington. The result is that as we enter a new era of opportunity for nuclear energy, not enough is being done to fix the underlying systemic issues. Instead, the focus remains on doing more of what has not worked in the past, which is using subsidies to hide the costs of a broken system and to bias against new technological approaches to commercial nuclear power.

The fact is that a properly functioning system would provide the best opportunity to work out all these cost inflators over time. The delays and cost overruns, which you'd find with most any new commercial endeavor, should diminish as the industry grows. As backlogs are created by new orders, suppliers will invest to expand capacity. And we know this will happen. For example, in response to anticipated new construction in the mid-2000s, Japan Steel Works (JSW) expanded its capacity to produce the large forgings used to manufacture reactor components. The company tripled its yearly capacity for reactor pressure vessels and related major parts from just four in 2007 to twelve by early 2011.[10] JSW holds 80 percent of the global market for these forgings. There are many other examples from those years of private money flowing toward commercial infrastructure expansion. But nearly all of it came crashing down when the anticipated new nuclear construction never materialized.

To be sustainable, investment must be made in response to organic market forces and not to markets manufactured in Washington. We see private money beginning to flow toward nuclear again today, with companies announcing plans to provide expanded uranium enrichment, mining, manufacturing, and used-fuel services. If the government could just get out of the way, this growth in capacity would naturally meet demand and moderate some of the inflationary pressures that have driven up costs in the past.

But this is much easier said than done. Getting government out of the way will be difficult—and it may even be impossible without reforms to allow for the emergence of a system that is very different from the one that exists today. Unfortunately, many policymakers don't think about what nuclear could (or even should) cost, and instead they argue that it's just too expensive, and that America should therefore pursue alternative low-carbon sources of energy. But that's a hollow argument that lacks appreciation for the broader context of energy policy, energy inflation, and rising construction costs.

In liberalized energy markets, this wouldn't matter. Advocates for wind and solar could invest in what they believe in, and the same would be true for those who support nuclear energy (or hydrocarbons, for that matter). The market does a good job of efficiently allocating resources to the best and highest use. The problem is that energy markets are far from liberalized and are becoming tighter every day as government distortions continue to increase.

Because of this, wind and solar advocates have an incentive not only to paint nuclear in the most expensive terms possible—which is easy, given recent history—but also to use that line of attack to undermine efforts that would allow the industry to reduce costs. The idea is to create the perception that nuclear is expensive relative to renewables and then do whatever is possible to keep those prices (or the perception of those prices) high.

The problem for wind and solar advocates, however, is that cost overruns are not specific to the nuclear industry. And unlike nuclear, which can clearly connect a significant portion of its high costs to bad government policy (at least to date), wind and solar actually *are* expensive.

But the point of this book is not to condemn or criticize any specific energy source, including renewables, or even to advocate for any other energy source per se. The point is to argue for a policy and regulatory environment that allows for all energy sources to compete fairly. Wind and solar power may have a role in America's energy mix, but those technologies alone are not ready or able to affordably and reliably power the country. They are expensive, intermittent, and inappropriate for broad swaths of the United States. Even environmental activists are beginning to oppose wind projects because they kill birds, despoil landscapes, and ruin scenic views.[11]

And renewables remain expensive, despite the rhetoric we often hear. The costs of offshore wind projects, for example, increased by 57 percent from 2021 to 2023 before the Inflation Reduction Act bonus credits were accounted for.[12]

A major problem with wind is its intermittence—it produces electricity only about a third of the time. This means that power plants are needed to provide electricity when the wind is not blowing. If we're going to rely on wind, the additional costs associated with power-generating capacity that is needed when the wind is not blowing should be factored in. What's more, the projected life expectancy of wind turbines is just twenty to twenty-five years,[13] whereas nuclear plants can produce power for eighty years or more. This must be taken into account when considering not only costs but also long-term business plans.

Solar energy projects are also running into trouble. Like wind, solar is intermittent: it produces electricity only when the sun is shining. As more residential solar capacity is installed, it reduces the net load during the middle of the day, when solar generation is at its peak. But as the sun sets and solar generation rapidly decreases, the net load curve spikes, resulting in what has been named the "duck tail curve."

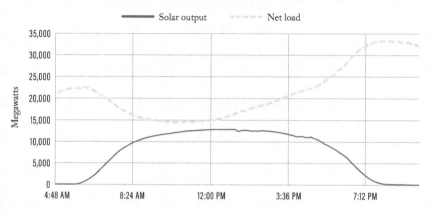

SOURCE: California ISO.

The duck tail curve presents several challenges for grid operators, including the possibility of overloading the grid and the need to rapidly ramp up energy that can be quickly dispatched to make up for grid shortages during peak hours.[14]

As long as the lights turn on and the AC kicks in, people don't need to think about where their electricity comes from, but they do care how much they pay for it. The economic consequences of forcing intermittent energy sources onto the grid become clear in the higher electricity prices paid by residents in states that mandate them. Eighteen states plus DC have a renewable portfolio standard (RPS) of 35 percent or greater by 2030. Residents in these states pay an average of 21.1 cents per kWh.[15] In contrast, residents in the thirteen states with no RPS pay 13.5 cents per kWh. In other words, residents in states where government heavily influences the energy mix face electricity bills that are 56 percent higher, while also bearing the implicit costs of decreased grid reliability. The freer the energy market, the cheaper the energy.

As wind and solar are deployed more broadly, other problems have also arisen. For example, nuclear plants were being put out of business in some markets because they cannot compete with heavily subsidized wind power that enjoys state mandates and special tax treatment. How is that a problem for wind and solar advocates? It's not, but it is a real problem for American families and businesses because these policies will result in higher electricity prices (certainly once the subsidies go away) and degraded reliability. Policymakers have tried to address this by offering nuclear power operators the same tax subsidies that wind and solar enjoy. But the way to counteract the consequences of existing subsidies is not to layer on additional subsidies. This may help an unsubsidized power source compete against those that are subsidized, but it also imposes substantial economic distortions that will result in inefficiency and waste over time.

In California, where residential and commercial solar is plentiful, the price of electricity is greatly reduced and can even turn negative at some points in the day.[16] While that may sound like a good thing for your energy bill, it prompts baseload generators to shut down and restart, which reduces efficiency and ends up raising the cost of electricity when it is most in demand. The result is higher electricity prices overall, which is again demonstrated by California, which has some of the most expensive electricity in the country.[17]

The intermittent nature of both solar and wind is important to the overall economics of energy. We've established that solar power is inherently intermittent, but it's also unpredictable. There are implicit costs to managing such variability in a power grid where supply must meet demand 24/7. Whether solar produces more or less than consumer demand, a measure must be taken to rectify the imbalance. If it produces too much energy, it must be curtailed to avoid overloading the grid. If (when) it doesn't produce enough, another dispatchable source must quickly make up the difference. A cloud that drifts over the sun can reduce solar output by 75 to 90 percent.[18] These curtailment and intervention operation fees are a significant economic burden for solar. In addition, transmission costs for intermittent energy often require major investment because new solar plants and wind farms are typically located far outside city centers, and in many cases, in totally different regions.

Though the promise of advanced battery technology is often touted as a viable method of overcoming the intermittency problem, such technology is not currently available at competitive prices (if at all). The need to rapidly ramp up other power sources to accommodate solar's variability leads to increased wear and tear on equipment, reduced efficiency, and higher maintenance costs for these backup systems. Worse, the subsidies that keep these intermittent power sources artificially economical undermine the market incentives to develop the technology that could allow them to be competitive without being subsidized. The right policy is to get rid of the original subsidies and allow all energy sources to compete on a level playing field.

Every large intermittent energy project like wind and solar will need some kind of dispatchable energy source to back it up, and if the goal is to deeply reduce carbon emissions, that energy will have to be nuclear. Given the low cost of operating a nuclear plant once it is built, why would any business conclude that wind or solar farms should be constructed at all? The nuclear plant, once built, will produce electricity regardless of whether the sun shines or the wind blows.

This could explain why many opponents of nuclear power are committed to renewable portfolio standards, which twenty-eight states and the District of Columbia have imposed. These standards dictate that a certain percentage of energy production capacity must come from wind, solar, or some other renewable source. This forces energy producers to invest in wind and solar even when such investment is not economically rational. California's RPS, which aims to achieve 60 percent renewable by 2030, explicitly excludes new nuclear energy as renewable, while not discriminating against wind or solar.[19]

The good news is that the untenable nature of these standards has begun to change policy when it comes to nuclear. For example, Kentucky, Montana, West Virginia, and Wisconsin have fully repealed restrictions on the construction of new nuclear facilities.[20] While these new broader standards are an improvement, they still fall short of optimal energy policy.

Instead of mandating how energy is produced, the government should simply set the regulatory framework and allow the market to dictate how best to meet America's energy needs. If wind and solar are competitive, they will succeed. The same holds true for nuclear.

EFFICIENCY MANDATES DEVALUE THE ROLE OF NUCLEAR ENERGY

Even critics of nuclear energy acknowledge that wind and solar sources alone will not meet America's growing energy demand. It is a tacit acknowledgment that green policies will result in inadequate energy production. Rather than recognize the underlying issues with these policies, though, they assert that any shortfalls in supply can best be met by increased efficiency. When they recognize further that the costs to achieve the sorts of efficiencies that they want are too high for people to pay for voluntarily, they promote mandatory energy reductions to assure it happens. Of course, they don't call them mandatory energy reductions, or energy rationing, which is what they are; they call them efficiency standards. We can see these so-called standards being applied throughout the economy in everything from dishwashers to commercial buildings.

Make no mistake, these mandates have nothing to do with saving consumers money and everything to do with imposing energy rationing to obscure the real impact of the green agenda. Such draconian measures are dangerous and will ultimately result in dire economic and societal consequences as families and businesses are left with fewer options and will eventually be forced to ration energy.

Mandatory efficiency requirements will never work because they often raise the prices of consumer goods and force engineering in directions that technology is not ready to support. In economics terms, they force a suboptimal use of scarce resources by artificially overvaluing certain things (in this case, efficiency) over others. Over time, the result is usually lower productivity, less efficient technological innovation, and higher prices. This not only affects everyday lives (toilets don't effectively flush, and washing machines don't effectively wash) but also can have broader technological ramifications. For example, the new National Highway Traffic Safety Administration (NHTSA) proposes standards to force car manufacturers to focus their research and development resources on the electric vehicles that Washington bureaucrats and politicians prefer instead of on revolutionary transportation technologies that meet consumer needs and preferences. NHTSA wants the new standard for passenger cars to be nearly 70 miles per gallon by model year 2032, which they acknowledge will be met only if two-thirds of American cars are electric.[21]

None of this is to downplay the importance of energy efficiency. Energy resources are precious, and society benefits by their conservation. The value of efficiency mandates is dubious, however. People's interests are served by efficiency, and they will pursue it where it most benefits them. But they are not served *solely* by efficiency. Personal preference and product performance also matter.

Efficiency mandates essentially assign a moral value to energy conservation rather than recognize it as an economic variable to be considered in decision-making. As a resource becomes less available, its price will rise. This sends a very direct signal to consumers that they need to either use less of that resource—in other words, conserve it—or find alternatives. Such an approach invites capital to move toward innovations that can fill energy voids. Sometimes these innovations will come in the form of efficiencies, but at other times, they will come in the form of new energy resources.

This is where nuclear comes in. It can fill supply gaps organically, instead of using policy to artificially balance supply and demand. In other words, the energy sources that make the most economic sense while also meeting the preferences of families and businesses will be the ones that move forward. The result will be a mix of energy conservation and innovation that yields the best results for the most people. But this will happen only if government steps away.

THE PERPETUAL MEDIOCRITY LOOP

When government bureaucrats and politicians use the force of law to determine how energy is produced, society loses the benefit of the investment and innovation that would have occurred without subsidies and mandates. For example, between 2016 and 2022, according to the Energy Information Administration, the US government doled out $183 billion to energy companies. These are not imaginary dollars that grew on money trees. They are real dollars, produced by American businesses and workers, that the government taxed out of the productive economy to redistribute as it deemed appropriate.

Now imagine what American businesses and entrepreneurs could have produced with that $183 billion, or what preferences American families could have secured if they had directed that money to its highest and best use.[22] We don't know what could have been, but it probably would have been more than a bunch of subsidy-dependent wind turbines, solar panels, carbon capture schemes, and endless other failed energy projects. There would certainly have been some failures along the way, but there would also have been some tremendous successes.

The irony is that if government would stop mandating and subsidizing intermittent energy, there would be a tremendous incentive to find greater cost efficiencies with those technologies and to innovate new technologies that could possibly compete with nuclear and hydrocarbons. But subsidies and mandates distort that process, essentially killing innovation.

Take, for example, the massive wind farms that attempt to duplicate the model of high-output centralized power stations. It may be that a decentralized model—where households or neighborhoods have their own energy sources—would work better for some of these technologies, while more centralized models would work for others. But because the government attempts to funnel money in one direction, investments in these technologies tend to follow the mandates rather than consumer preferences or economic value. The result is outcomes that are curated to meet temporary political fads rather than sustainable economic order.

The same is true for nuclear energy. Many in the pronuclear community are excited about the broad bipartisan support and massive taxpayer funds that nuclear is currently enjoying. If such foundations were the key to building successful industries, then there would be much to be excited about. The problem is that they are not. This approach rests on an underlying presumption that the transition from a hydrocarbon-based energy economy to something else can be forced by the government. It cannot.

Frankly, it's ridiculous to think that a perfectly good system of energy production can essentially be shuttered and replaced by something else not only without massive economic cost but with economic gain. The idea essentially disregards the fundamental basis on which every successful modern economy is based, which is that competition for meeting consumer demand, combined with the drive toward profits, is what motivates firms to seek better and more efficient ways to produce products and compete in the market. In other words, the natural incentive to develop new energy sources that can displace established ones is that it will make you rich. If you could produce energy more cheaply and efficiently using windmills and solar panels, the government wouldn't have to force people to buy them. They just would. As it is, subsidized technologies get stuck in perpetual mediocrity as firms spend more time and resources on profiting from government largesse than they do on increasing their ability to compete in the market.

As energy subsidies become more widespread, something even more nefarious takes place, which is the economic destruction of competitive and viable energy sources. That happens when the government subsidizes uneconomic energy sources at such high levels that they can undersell their unsubsidized competitors or even force the premature shutdown of perfectly good and profitable energy infrastructure through regulation or mandates. When this happens at the systemic level, which we are beginning to see in the US now, the result is a deterioration of energy reliability and inflated prices as secure and affordable energy sources are destroyed to make room for suboptimal sources that government prefers.

If policymakers are forcing a transition that is not happening organically, we must acknowledge it for what it is, which is a mandate to destroy, either for the sake of destruction or to advance the agenda of a select group of powerful elites. The simple fact is that what the people want destroyed, economically speaking, will nearly always be destroyed without government intervention. Have you tried to purchase a VCR lately?

Essentially, those advocating a forced energy transition are asking us to forget what French economist Frédéric Bastiat taught us with the broken window fallacy from his 1850 essay "That Which Is Seen, and That Which Is Not Seen."[23] In that historic piece, Bastiat wrote:

Whence we arrive at this unexpected conclusion: "Society loses the value of things which are uselessly destroyed;" and we must assent to a maxim which will make the hair of protectionists stand on end—To break, to spoil, to waste, is not to encourage national labor; or, more briefly, "destruction is not profit." What will you say, Moniteur Industriel? what will you say, disciples of good M.F. Chamans, who has calculated with so much precision how much trade would gain by the burning of Paris, from the number of houses it would be necessary to rebuild? I am sorry to disturb these ingenious calculations, as far as their spirit has been introduced into our legislation; but I beg him to begin them again, by taking into the account that which is not seen, and placing it alongside of that which is seen.

Politicians and bureaucrats will proffer all sorts of justifications for their approach. They will claim cost savings, environmental benefits, national security advantages, and so on and so forth. And with the strategic use of subsidies, rhetoric, and other political tricks, those seen returns may well result in popular and political support in the near term. But ultimately, the hard truth of the unseen costs that Bastiat described will be revealed, and support will quickly unravel.

That's why building any industry around a political consensus will fail. Unlike economic truth, politics is fleeting. This will be especially true for a forced transition that promises lower-cost energy and environmental gains and is described in terms of an existential fight against global warming. But regardless of where you stand on the science of human-caused global warming, the undeniable truth is that a forced energy transition like the one currently being imposed by Washington will never work.

CHAPTER 6
AN INDUSTRY GASLIGHTED

Commercial nuclear power has been around since the late 1950s. The nuclear industry demonstrably knows how to build and operate reactors safely. In fact, the first ten commercial reactors built have the same safety record as the last ten, which is the same safety record as every one in between: no one has ever died or even been sickened by a radioactive event related to commercial nuclear power in the United States.

Despite this stellar record, the industry has been subject to an ever-increasing regulatory burden, which has led to massive cost escalations for constructing nuclear power plants. But this wasn't always the case. The United States used to build reactors in reasonable amounts of time and for reasonable amounts of money. In fact, between 1954 and 1967, the industry controlled and in some cases reduced the costs associated with building its first thirty-two nuclear power plants. The next forty-eight that came online before the Three Mile Island incident were progressively more expensive to build, with a total increase of 190 percent. This was unnecessary and unfortunate, but it was arguably manageable. The fifty-one reactors completed after Three Mile Island, however, experienced further price increases of between 50 and 200 percent.[1] And this is what killed the promise of American nuclear power for nearly three decades.

The United States experienced the highest cost increases over that general span of time, but France, Japan, and West Germany had a similar outcome. South Korea, however, did not. In fact, for the nine foreign-designed reactors it built between 1972 and 1993, South Korea achieved a cost *reduction* of 25 percent. And for the nineteen domestically designed reactors built between 1989 and 2008, the reduction was 13 percent.[2]

This tells us that we shouldn't accept the idea that the cost to build new nuclear plants necessarily increases over time. The US was able to control costs as part of its initial nuclear program, and South Korea has done it more recently. Yet we're constantly told that nuclear energy is too expensive.

Essentially, the industry has become the focus of a multi-decade gaslighting campaign. Antinuclear activists have generated unwarranted fear around the safety of nuclear energy and used that to lobby for additional regulation. This has contributed to a massive increase in prices, which the same activists then use to argue that nuclear energy is too expensive. At the same time, they have used the subsidized cost of wind and solar to claim that they are more affordable than nuclear energy and therefore should continue to enjoy taxpayer support. Meanwhile, advocates for nuclear energy have fallen into the trap of arguing that the industry also needs subsidies to either compete with subsidized renewable energy or restart the industry.

How did this happen?

ACTIVISTS GONE WILD

Antinuclear groups used both legal intervention and civil disobedience to impede construction of new power plants and hamper the operations of existing units. They challenged 73 percent of the nuclear license applications filed between 1970 and 1972 and formed a group called Consolidated National Interveners for the specific purpose of disrupting hearings of the Atomic Energy Commission.[3]

Much of the antinuclear litigation of the 1970s was encouraged by factions within the government.[4] Today, activist organizations determined to force the closure of nuclear power plants, such as Mothers for Peace, continue to use the legal process to hinder the industry. Though public perception is shifting in favor of nuclear at the national level, local activists continue to successfully shut down perfectly good plants. For example, persistent protests by so-called environmentalists and activists significantly contributed to the closure of the Vermont Yankee power plant in 2014.[5]

In fact, thirteen nuclear reactors were prematurely shut down between 2013 and 2022.[6] While not all of these closures were due to local opposition, the fact remains that antinuclear activism, combined with market-distorting energy policy, contributed to most. Over the past decade, the United States has lost more than 10,000 MW of reliable and clean power.

PREMATURELY RETIRED REACTORS				
REACTOR	STATE	NET CAPACITY (MWe)	YEAR SHUT DOWN	YEARS PREMATURE (assuming 60-year lifespan)
Crystal River 3	FL	860	2013	24
San Onofre 2	CA	1,070	2013	30
San Onofre 3	CA	1,080	2013	31
Kewaunee	WI	566	2013	21
Vermont Yankee	VT	605	2014	18
Fort Calhoun	NE	482	2016	17
Oyster Creek 1	NJ	619	2018	11
Pilgrim 1	MA	677	2019	13
Three Mile Island 1	PA	819	2019	15
Indian Point 2	NY	998	2020	14
Duane Arnold	IA	601	2020	14
Indian Point 3	NY	1,030	2021	15
Palisades	MI	805	2022	9
		10,212		

SOURCE: World Nuclear Association.

Activists went well beyond simply challenging nuclear power in the courts. On numerous occasions, demonstrators occupied construction sites, causing delays. For instance, in May 1977, the Clamshell Alliance led a protest that resulted in the arrest of more than fourteen hundred people for trespassing at the Seabrook plant site in New Hampshire.[7] In California in 1981, the Abalone Alliance adopted similar tactics and frequently blocked the gates of the Diablo Canyon power plant.[8]

The antinuclear movement scored a watershed victory in 1971, when a federal appeals court ruled that the construction and operating permits for a nuclear power plant in Maryland violated the National Environmental Policy Act of 1969.[9] (NEPA involves an environmental review, which can be a years-long quest through federal bureaucracy.)[10] As a result, utilities were required to hold public hearings before they could get permits to start projects. This decision created a major opening in the regulatory process that antinuclear activists were able to exploit.

CHANGING THE ECONOMICS OF NUCLEAR POWER

The commercial nuclear industry emerged in its early days from strong cooperation between the public and private sectors. Some have argued that this early support from the US government constituted a broad subsidization without which the nuclear industry could never have been established.[11] They often further argue that because of that, similar support for renewables is only fair.

The first part of this argument is true. Commercial nuclear power would certainly not have been available when it was, if at all, had the government not developed the underlying technology for national security purposes. Furthermore, Washington's conclusion that a successful commercial sector could support the national security enterprise incentivized additional (arguably justifiable) interventions during early commercialization. There is no analogous context for renewables. The argument's second part holds no merit, at least not today. The fact is that renewables have been subsidized for decades. So even if there was an argument that new industries should be subsidized, that time has long passed for wind, solar, and nuclear.

Legitimate or not, this early public–private cooperation put the nuclear industry on a trajectory that made it overly vulnerable to government intervention and locked it into technological approaches like the light-water reactors that primarily define the industry and policy today. As a result, the industry did well so long as public and political support for it was high. But the antinuclear movement's success in undermining this support led to a broad rethinking of how government institutions were set up to interact with the industry in ways that ultimately created unnecessary impediments.

For example, in Congress, the Joint Committee on Atomic Energy was disbanded in 1977, and oversight responsibility for nuclear activities was transferred to multiple other committees. This led to decentralized oversight and a weakening of nuclear policy overall. It also provided additional avenues for antinuclear lobbyists to influence Congress.

In the executive branch, the Atomic Energy Commission, which both advocated for and oversaw the nation's nuclear activities, was replaced in 1975 by the Nuclear Regulatory Commission (NRC), whose sole function was to regulate the nuclear industry. Advocacy responsibilities went to the Department of Energy. Although the Department of Energy does a good job at developing nuclear energy programs and spending taxpayer money to support them, there is little evidence that its "advocacy" has resulted in an independent, innovative, and economically sustainable industry.

In addition, the role of the judiciary cannot be overemphasized. Congress's loss of enthusiasm for nuclear energy led to more aggressive regulation, and because jurisdiction over nuclear issues was divided among multiple committees, there was no unified congressional stance. One result was an expansion of the regulatory environment that persists today.

As of June 2024, the NRC listed over eighty sources of regulation, including over two thousand pages of laws, treaties, statutes, authorizations, executive orders, and other documents.[12] Policy is statutorily set by legislation like the Atomic Energy Act of 1954 and the Energy Reorganization Act of 1974, but nuclear activities in the United States must also comply with such laws as the Inspector General Act of 1978, the Clean Air Act of 1977, the Federal Water Pollution Control Act of 1972, and the National Environmental Policy Act of 1969.

This created numerous opportunities for antinuclear groups to file noncompliance suits. Regardless of whether the groups' concerns were legitimate, regulators often responded with additional mandates, which were very easy to establish. A regulator could compel a change in plant design, for example, simply by deciding that such a change would add substantially to public health or safety. The problem was that NRC statutes did not define "substantial." Because the interpretation of NRC regulations was left to the discretion of individual technical reviewers, each license application often resulted in its own unique requirements.

This inconsistency increased costs, further souring Congress on nuclear power and leading to an endless spiral of legislation, regulation, and still more added costs. Between 1975 and 1983, 430 lawsuits were brought against the NRC, leading to 2,349 proposed rules and regulations—each of which required an industry response. The additional and unexpected controls created industrywide uncertainty and raised questions about the long-term economics of nuclear power.[13]

This was all done by the NRC without adequate information. The commission recognized as early as 1974 that it was issuing regulations without sufficient risk-assessment training or cost considerations. It didn't even have a program to train employees in how to conduct a review using NRC guidance. Yet the commission continued to issue regulation after regulation.[14]

At the same time, state and local governments expanded their oversight functions. States often claimed influence over construction and operations permits, as well as environmental regulation. For example, while the Federal Water Pollution Control Act (as amended by the Clean Water Act of 1977), the Clean Air Act, and the Solid Waste Disposal Act mandated that states enforce minimal federal environmental standards, many states choose to adopt additional regulations.[15] Environmental standards that varied from jurisdiction to jurisdiction imposed additional costs and opened additional avenues for antinuclear activists to exploit.

Today, many states exercise significant authority over the location and construction of nuclear reactors.[16] Some jurisdictions even have moratoria on new nuclear construction.[17] For example, California will not allow any further construction of nuclear plants until both the California Energy Commission and the federal government approve a method of disposing of nuclear waste.[18] Most states that limit construction use some variation of this theme. Public commissions and referenda can impose additional restrictions.

The shifting regulatory environment gave rise to additional reviews from numerous public institutions. Once permits were obtained, additional design changes were often mandated—even during construction. Between 1966 and 1970, this inefficient process increased the time required to build a nuclear power plant by 42 percent (from 86 months to 122). From 1974 to 1984, the average construction delay was nearly 40 months, and between 1956 and 1979, the average time for construction permit review increased fourfold.[19] The average time required to bring a plant online from the order date increased from four years to fourteen during a similar period.[20] This significantly increased both the cost of each plant and the risks to the investors financing these projects. As the need for electricity increased, lengthy delays further undermined public confidence in the viability of nuclear power.

During the 1970s, regulatory mandates also drastically increased the quantity of materials required to build a plant. Steel requirements increased by 41 percent, concrete by 27 percent, piping by 50 percent, and electrical cable by 36 percent.[21] Even though experience demonstrated that these increases were not needed to maintain safe operations, regulatory relief never followed. In some instances, builders even added safety features that were not mandated in hopes of preempting further stoppages.

More inspections were required, and delays often resulted from a lack of personnel to carry them out. Workers had to spend inordinate amounts of time waiting for inspections rather than building projects. The changing construction specifications also led to additional complexity and mistakes, which created further delays. Even after construction was complete, the delays often continued—sometimes at a cost of up to $1 million per day.[22]

The additional regulation also caused operations and maintenance (O and M) costs to increase. For example, from 1981 to 1988, O and M costs increased by 80 percent, and 30 to 60 percent of this increase was the direct result of NRC regulation. In total, almost 75 percent of working hours were lost to dealing with NRC bureaucracy. In any given week, this included eleven hours lost due to unavailable tools and materials, eight hours lost to work area overcrowding and coordination issues, and just under six hours lost to redoing work.[23]

In the high-interest 1970s, long delays significantly increased project costs as rising interest payments drove up the cost of capital. High inflation impacted the costs of materials. Sometimes, plants were completed, or nearly so, and ready to start producing electricity but were not allowed to begin operations for one regulatory reason or another. In the 1980s, the Shoreham nuclear plant on Long Island was completely built but never used because extremists succeeded in scaring the public and political leaders.

OVERREGULATION LEADS TO A DECLINING INDUSTRY

From 1971 to 1980, regulation increased the overall costs of constructing a nuclear power plant fourfold.[24] That might have been justified if it had been rooted in scientific and technical analysis. Regrettably, it was largely driven by antinuclear activists, agenda-driven politicians, activist regulators, and unsubstantiated public fear.[25] A fascinating 1990 study by Dr. Bernard Cohen, who was widely published and won multiple awards for his nuclear physics research, lays forth in excruciating detail the price escalations that occurred during the 1970s and 1980s. Dr. Cohen found that the typical cost for a nuclear plant in the 1970s, adjusted for 2024 dollars, was around $1.2 billion ($170 million nominal). By 1983, those costs had escalated to around $5.4 billion ($1.7 billion nominal). And by the late 1980s, prices had risen to as high as $13.5 billion ($5 billion nominal). This was clearly not sustainable.

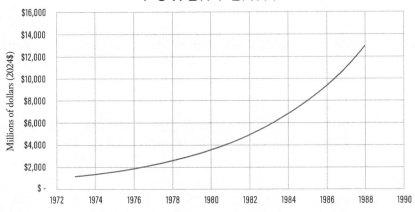

SOURCE: Data from Bernard L. Cohen, *The Nuclear Energy Option* (New York: Plenum Press, 1990).

The antinuclear sentiment propagated by activists didn't end with nuclear power plants. The US government even banned entire commercial technologies outright. In 1977, for example, following a pause imposed by Gerald Ford, Jimmy Carter dealt the industry one of its greatest setbacks by issuing Presidential Directive 8 (PD-8), which forbade reprocessing (recycling) used nuclear fuel in the United States.[26] The policy then informed the 1982 Nuclear Waste Policy Act, as amended, which made the federal government solely responsible for managing and disposing of all commercial spent nuclear fuel, and also dictated, by virtue of later amendment, that spent fuel be disposed of in a permanent waste repository at Yucca Mountain in Nevada. Though Ronald Reagan later lifted Carter's commercial reprocessing ban, the decision to make Washington responsible for nuclear waste management, along with a legal technology determination, essentially killed the incentive to reprocess spent nuclear fuel.

Recycling spent nuclear fuel could, depending on the technology applied, reduce the volume of high-level nuclear waste and recover vast amounts of energy that remain in "spent" nuclear fuel that has gone through a reactor. It would also create the opportunity to extract other useful components, such as rare earth elements and medically valuable isotopes.[27] Currently, only about 5 percent of the energy is used per volume of fuel. The US does not recycle nuclear fuel, but France, China, and Russia are safely using recycling technology.[28]

With recycling in place, the US nuclear industry could finally move away from relying on the all-but-defunct Yucca Mountain repository. It would allow for a more reasonable mixed approach to nuclear waste management and disposal, which would likely include some combination of permanent geological storage, interim storage, recycling, and new technologies. But this will never happen under current policy, which not only disincentivizes commercial reprocessing but also is susceptible to the same cost inflators that have hampered the construction of new power plants.

That's one reason why it won't be easy to establish economically viable commercial recycling in the US. President Carter's unilateral ban had a chilling effect on the domestic nuclear industry because it forced suppliers to discontinue their activities at the cost of hundreds of millions of dollars. One industry group had invested approximately $500 million in a Barnwell, South Carolina, project that never became operational.

It must be noted that there is no guarantee that any of these facilities would have been economically viable. The fact is that they were being built within the context of a nuclear industry whose history was more closely tied to meeting national objectives than economic ones. However, the point still stands that America's nuclear industry has been subject to incoherent government policies throughout its existence, and this is just one more example.[29]

With overregulation driving up the cost of nuclear power and the government banning critical commercial technologies, the incentive to build new nuclear plants all but died. From the early 1950s through 1974, 231 nuclear power plants were ordered. Another 15 had been ordered by 1977.[30] But until the reactors in South Carolina (which were never completed) and Georgia (Vogtle) were requisitioned around 2010, no new orders had been placed since the mid-1970s.

Not only did orders stop, but previously ordered plants were cancelled. Of the 246 plants ordered prior to 1977, only 92 still operate today. Some were never built, while others were shut down early. In many cases, construction was stopped after substantial investments had already been made, resulting in billions of dollars in losses. For example, the Cherokee plant in South Carolina was cancelled in 1982 after over $600 million had been invested. In 1983, a group of three utilities cancelled the Zimmer plant in Ohio after investing $1.8 billion. In total, over $35 billion in today's dollars has been spent since 1972 on nuclear plants that were never completed.[31]

The result is that the US commercial nuclear industry declined significantly and rapidly. Meanwhile, other nations, such as South Korea, Japan, France, Russia, and China, continued moving forward and are now well positioned to lead the global resurgence in peaceful nuclear power.

RECENT REGULATORY REFORMS ARE INADEQUATE

The problem of overregulation is nothing new and has generally been recognized as a serious barrier to building new nuclear plants. Efforts by Congress and various administrations to clear some of the regulatory hurdles to construction have been going on for decades.[32] For example, the Energy Policy Act of 1992 allows utilities to combine their construction and operations licenses, which should have streamlined much of the regulatory process.

The problem, however, is that efforts to reform the system have not really worked. While they have solved some issues, the time and costs required to build new plants are prohibitive. The Energy Policy Act of 2005 attempted to mitigate the risks by allocating money to protect plants under construction from regulatory delays.[33] But ultimately, this effort only served to transfer the costs of overregulation to taxpayers and did nothing to fix the underlying issues.

Many congressional leaders continue to offer legislative reform packages to relieve the nuclear industry of the bureaucratic burden imposed by federal regulation. Congressman Byron Donalds, for example, has introduced countless pieces of legislation to help modernize how the US regulates nuclear power.[34] Other recent examples include the Nuclear Energy Innovation and Modernization Act, the Atomic Energy Advancement Act, and the ADVANCE Act—all of which aim to make it easier and faster to build and operate new nuclear power plants.[35] Each of these bills is designed to simplify the rules and processes for approving new nuclear commercial activities, speed up the decision-making for licensing, and ensure that environmental reviews are done quickly and efficiently.

It remains to be seen if these most recent efforts will ultimately help, but past ones did little for long-term predictability. Instead, they provided just enough regulatory relief and taxpayer support to ensure that a small number of plants (two, to be exact) would be built. There was no nuclear renaissance. America is still in the same position it's been in for decades.

The near death of the US nuclear energy industry has harmed both investors and consumers. At the end of the day, ratepayers are the ones who pay for the increased costs of generating electricity. And by reducing nuclear's role in America's energy portfolio, activists have limited the choices available to producers and consumers. Limiting choice has two inevitable results: higher prices and lower quality.

Today, utilities are moving toward less reliable and more expensive power sources like wind and solar while still relying heavily on natural gas, which has a lot of benefits but historically comes with significant price volatility. The result is an emerging power grid that is more expensive, less reliable, and more dependent on foreign suppliers. And this is all in the name of reducing CO_2.

Imagine what America's energy profile would look like had the nuclear industry been allowed to stay on the trajectory it set for itself in its early years. What if all of the 246 plants ordered in the 1970s and 1980s, plus the nearly 30 more that were considered during the mid-2000s, had been built? Not only would we have had well over double the nuclear capacity that we have today, but it's likely that many more plants would have been built as well.

More importantly, think what power prices would be like. If we had taken a different approach, it's not hard to imagine a world where electricity might be way less expensive than it is today and far less subject to the massive price swings we so often see.

And let's take our imagination a step further. If we accept for a moment that carbon reductions are desirable, let's think about what the world would look like if the government had simply done nothing. What if it had stepped aside and allowed energy producers and consumers to decide the best ways to produce and consume electricity? Well, it goes without saying that the move toward a less carbon-intensive electricity grid would be far less destabilizing if nuclear had been allowed to stay on its natural trajectory. And when coupled with the advances made with natural gas development and highly efficient coal technology (which emits 26 percent less CO_2 than the average coal plant), the US grid would have been far greener and completely driven by the market.[36] While we still wouldn't have achieved the net-zero objective that CO_2 alarmists claim to want, our CO_2 intensity would have been substantially diminished, and politicians would have had a far more difficult time wreaking havoc on the grid with their myopic energy industrial policy.

It's not too late.

CHAPTER 7
GOVERNMENT CAN'T SUBSIDIZE AN INDUSTRY INTO SUCCESS

It doesn't take long for any conversation around energy subsidies to descend into a debate over which sources are most subsidized. But this is pointless. Like almost every conversation about energy policy, this one is defined as much by spin, selective definitions, conflation, and purposeful misdirection as it is by facts.

The truth is that nearly all major categories of energy sources have received some sort of subsidy at some point. A subsidy in this sense is any public policy that is used to bias the market in favor of a specific energy source *or* any taxpayer-funded payment (or the equivalent) received by a producer or buyer to reduce the price of a type of energy or energy-related product to help it succeed commercially. (Any policy that artificially inflates the price of a specific energy source would be considered a subsidy for that source's competitors.) Arguing for a larger subsidy for one energy source because a different source got something in the past is like a child trying to convince her parents that she deserves more ice cream to match her sister's serving. When it comes to subsidies, the only question should be, Is this a legitimate and effective policy tool on its own merit?

It's also important to note that not all government policies or programs that may help an industry are the same, and not all equate to a subsidy per se. The context of the interaction matters. For example, a government program to develop a technology that meets a national interest (even if the program has commercial implications) is not the same as a government program to commercialize a product. The former is not a subsidy because it's not an effort to manipulate markets in favor of one product over another. Similarly, just because a program was justified in the past doesn't mean that it can't evolve into an unwarranted subsidy in the future.

While politicians and special interests will offer an endless list of reasons to support subsidies, they all have at least one of two things in common: (1) they are the product of corruption or (2) they are used to cover up some economic inconvenience. In the first case, public officials use subsidies to funnel government largesse to themselves or their friends and associates, or to advance a political agenda. In the second case, subsidies are used to make an uneconomical product seem more affordable. For example, let's say a certain renewable energy product doesn't make economic sense but has strong political support. Subsidies can allow that product to be successful so long as they remain in place. Affordability can come from the consumer side as well. For instance, energy subsidies are commonly provided to low-income households that may not otherwise be able to afford their energy needs.

The problem with all this is that subsidies do not exist in a vacuum. They distort all the economic decision-making around the subsidized product and the markets in which it exists. Subsidies impact where private investment goes, how consumers interact with products, how industries evolve over time, how government policy evolves, and how businesses compete. The longer the subsidies stay in place, the more impact they have. Over time, they can be very economically destructive.

When it comes to nuclear energy, bad policy has made success for the industry all but impossible. The underlying economics have been distorted not only by subsidies but also by antinuclear policy and regulatory bias, as we have seen. This has made the true costs of nuclear power impossible to ascertain. Still, we know that it can be far more affordable, as it was in the past. That tells us there must be underlying problems scaring away investment in this proven technology. Today, despite the clear and plentiful benefits of nuclear power, many advocates insist that subsidies are necessary to attract private investment.[1]

If we look at most of the policy proposals put forth in recent years, we might conclude that Washington thinks it can subsidize nuclear energy into commercial viability. Essentially, this was the premise behind the Energy Policy Act of 2005 (EPACT), which boosted the anticipated (but failed) nuclear renaissance of the mid-2000s, and the approach more recently embraced by the so-called Inflation Reduction Act (IRA) of 2022.

EPACT established subsidies adequate to support the construction of five or so nuclear reactors. Those subsidies were supposed to help the industry get off the ground so that private entities could begin new nuclear construction. Although the legislation did result in a series of building permit applications and even site work at two locations, it never led to a commercial nuclear rebirth. Indeed, since the law passed in 2005, only the two new Vogtle reactors have been built.

Instead of helping the industry reestablish itself, the law merely led to a proliferation of requests for additional taxpayer support. Since EPACT, Congress has introduced a veritable parade of legislation to broaden the federal government's support for the nuclear industry. Lawmakers have proposed increasing capital subsidies, using taxpayer money for such activities as workforce development and manufacturing improvements, empowering the Department of Energy to decide which technologies should move forward (and by implication, which should not), and directly bailing out struggling plants.

One of the basic problems with using subsidies to promote an industry is that it allows both government and industry to ignore the underlying problems that created the need for subsidies to begin with. The bailouts for struggling reactors are a perfect example of this. Those reactors, for the most part, weren't failing because of problems with the underlying economics. They were failing because they were forced to compete with highly subsidized intermittent energy sources like wind and solar. Rather than remove the subsidies for wind and solar—in other words, address the actual problem—politicians decided to subsidize nuclear. This approach perpetuates those structural issues and creates a cycle where industry becomes increasingly dependent on government support—and that is where nuclear power largely is today.

This is not to blame the nuclear industry. It has been unfairly treated and overregulated for decades. But the fact is that dependence dampens the incentive to create new technologies, finance approaches, and business models that would allow nuclear to succeed on its own merits. Why, after all, would anyone make those investments if government makes success all but impossible?

TIME FOR A DIFFERENT APPROACH

The good news is that over the years, companies have shown they are willing to invest in commercial nuclear energy despite the risks posed by the government. In fact, market conditions had already started inviting growth in the sector before the Inflation Reduction Act, even though virtually no reactor construction had begun. Instead, companies were expanding uranium enrichment and other fuel services, and numerous advanced reactor companies had emerged.

Some might argue that more growth is needed more rapidly and point to rising private investment following the Inflation Reduction Act as proof that subsidies work. But that would depend on how you define "work." If you mean that private investment will flow toward the sector and some reactors may be built as a result, then yes, the IRA likely will work. But if the goal is a competitive, innovative industry that is economically sustainable without ongoing taxpayer and policy support, then history tells us that it will not.

It bears repeating that federal intervention distorts private investment decisions. Only one of two things can happen: (1) the intervention encourages firms to misallocate scarce capital by investing in something that is otherwise less attractive than other options or (2) the intervention discounts the costs for an investment that a firm would have made anyway. Both cases will result in an inefficient marketplace—and eventually, a weaker overall industry, economy, and nation.

The question really comes down to how a nation derives strength. Does the mere existence of commercial nuclear plants strengthen the nation? Certainly not if they result in more expensive electricity. The same can be said for renewables. What does make the nation stronger is access to safe, clean, reliable, and affordable energy. So that should be the objective.

The next question, then, is whether nuclear energy can contribute to that objective. Sure, it can. The problem with nuclear energy is not safety, reliability, or even cleanliness. Experience demonstrates that the nuclear industry promotes each of these. The real challenge is affordability. But we know that nuclear can be affordable because it was in the past. What has caused those costs to skyrocket, and what can be done about it? Some people want to use subsidies. Though it ebbs and flows, taxpayer support has been consistent for decades, yet prices remain high. Then there is France, where the government owns much of the nuclear industry. Nonetheless, new reactors are as expensive there. There is simply scant evidence that subsidies work to bring down prices over time. The problem is too much government intervention into energy markets writ large.

That's because market-based pricing that reflects true supply and demand is an essential point of reference for any business. Government interventions that purposefully distort those prices undermine this critical economic foundation. The more pervasive the intervention, the bigger the problem. The energy industry is among the worst. For example, growing grid instability is not caused by energy producers deciding to shut down perfectly good coal and nuclear plants or to shoehorn in expensive and unreliable wind and solar. Instead, it is caused by government policies like production tax credits, renewable portfolio standards, mandates to reduce CO_2, costly regulations, and subsidies to wind and solar. To understand how pervasive these policies have become, consider how many states are living under renewable portfolio standards. The following page has a map showing the states where RPSs have been implemented.[2]

STATES WITH RENEWABLE PORTFOLIO STANDARDS

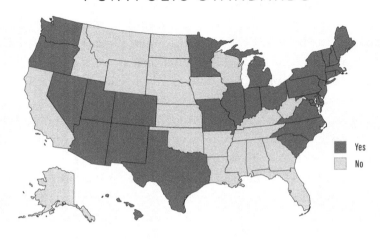

SOURCE: National Conference of State Legislatures.

Rather than acknowledge the negative impact of those policies, Washington uses subsidies and other tools to hide the true costs while blaming everyone else for the problem. As a result, the energy industry is never able to fully adjust to the market and spends its time consumed by the next dictate from Washington.

LEARNING FROM THE PAST

The energy crises in the 1970s prompted a significant expansion of publicly subsidized research and development for wind, solar, biofuel, and geothermal technology. In constant 2023 dollars, funding support for nearly all energy sources increased that decade, peaking in 1979 at around $10.7 billion.[3] Congress also passed a bevy of legislation in the late 1970s and 1980s designed to spur a renewable energy movement. For instance, the Energy Tax Act of 1978 promised residential tax credits for wind, up to 30 percent of the cost of technology for residents, and energy equipment expenditures and business incentives that allowed investors to receive tax credits of 10 percent in addition to the standard 10 percent investment tax credit.[4]

Subsequently, the Crude Oil Windfall Profit Tax Act of 1980, the Energy Policy Act of 1992, and the Economic Security and Recovery Act of 2001 all attempted to establish sustainable investments in and consumption of renewable energy. More recently, EPACT 2005 and the 2007 Energy Independence and Security Act required more agricultural-based renewable fuels and a host of other mandates and subsidies. The Inflation Reduction Act amplified that approach.

Despite all this, consumers have shown little faith in renewable energy's ability to meet demands. Of all energy production in the United States, only 8 percent comes from intermittent sources such as solar and wind, and that number has remained relatively flat over the past twenty years, increasing from 6 percent in 2004.[5] This is certainly not the result of too few subsidies. Since 2016, solar and wind have received over $55 billion. Solar alone has received more than coal, natural gas, nuclear, and geothermal combined.[6]

These legislative approaches will inevitably lead to higher costs for the consumer, and because everyone needs energy, that will disproportionately affect the poorest parts of the US population. Indeed, we've already seen this happening in Europe.

The European Union has faced significant challenges in its energy sector, with Germany closing its nuclear power plants and sanctions affecting the supply of natural gas from Russia. Despite these issues, the EU has continued to prioritize renewable energy mandates, further limiting the availability of hydrocarbon-based electricity. As a result, prices have skyrocketed to ten times their 2019 levels and are now stabilized at three times the pre-pandemic prices. Projections indicate that these elevated prices are likely to stick around. As my colleague Mario Loyola has pointed out, while this may have resulted in some decarbonization, it has also led to warnings of deindustrialization.[7]

Even though they are often presented as one of the most harmless subsidies, capital subsidies like loan guarantees are actually among the worst. These subsidies, under which the government guarantees bank loans for power projects, were originally promoted as a way to help move new clean energy sources toward market viability. Given the role played by organized political opposition and overzealous regulation in making the nuclear power industry uncompetitive, some limited near-term help to reduce government-imposed risk seemed appropriate.

In 2005, arguing in support of including nuclear energy as part of the Bush administration's energy program, Spencer Abraham, the former energy secretary, said, "I am not calling for massive ongoing subsidies to the nuclear industry, [but] I do believe some federal financial participation is in order to help defray a percentage of the high, first-time costs associated with new generation construction."[8] The same argument was made for other energy sources as well. But that is not how these government-backed loan programs evolved. In fact, they have become a pervasive, ongoing subsidy.

And it is a subsidy that should be ended. A 2009 exchange between Senator Richard Burr (R-NC) and Secretary of Energy Steven Chu during the latter's confirmation hearing was telling. Senator Burr suggested that the existing loan guarantee program was so poorly run that utilities were being forced to build reactors without the guarantees. Burr and Chu, embracing the subsidy-first mentality of modern US energy policy, concluded not that this demonstrates the market viability of nuclear power but that the subsidy program should be made more workable. They were inviting government dependence.

That's the problem with energy subsidies broadly and loan guarantees specifically: they distort normal market forces and encourage government dependence. The foundation laid during the EPACT 2005 debate has calcified into baseline policy today. According to one analysis, the total of DOE loan capacity amounts to $412 billion.[9]

By subsidizing a portion of the actual cost of a project through a loan guarantee or other subsidy, the government is distorting the allocation of resources by directing capital away from more competitive projects. This signals to industry (be it nuclear, wind, coal, natural gas, or anything else) that it does not have to be competitive. It reduces incentives to manage risk and be independent, innovative, and efficient. While a subsidy may be good for the near-term interests of the individual recipient, it is not good for consumers and taxpayers or for long-term competitiveness.

GOVERNMENT SHOULD SIMPLY STEP ASIDE

We know that the nuclear industry responds to market signals. The problems occur when those signals are defined more by government policy than by natural supply and demand. Too much government intervention results in firms evolving toward what government wants rather than what the market demands. This is the dynamic that played out with EPACT 2005, which provides an important learning opportunity for policymakers.

The industrial and educational sectors had already begun positioning themselves for additional nuclear business prior to EPACT. That was driven less by government policy to commercialize specific energy technologies and more by anticipated demand, the changing landscape of hydrocarbon prices (especially natural gas prices, which were spiking at the time[10]), and an emerging desire to reduce CO_2 emissions. In other words, the nuclear sector was responding to organic market signals.

At the same time, private companies had begun growing their workforces, enrichment and manufacturing facilities were expanding capacity, universities were increasing the size of their nuclear engineering programs, and the private sector was introducing craft-labor workforce programs. All this was in response to market demand for safe nuclear power—not because of a suite of government "incentives" set aside specifically for nuclear energy. Having said that, the government did introduce some regulatory reforms that were supposed to bring greater efficiency to the industry and generate increased interest in nuclear power.

The new approach allowed reactor companies to get preapproval for their designs through a design certification process.[11] This meant that those seeking to build plants could use a preapproved design in their construction applications. Site developers were also able to get early site permits (ESPs) that would iron out any site-specific safety or environmental obstacles that could slow down development.[12] While the design certification and ESP stages were independent of the application to build and operate a nuclear power plant, these reforms were intended to streamline the process overall.

The government also began to allow firms to apply to both construct and operate a nuclear plant with a single combined process.[13] Prior to this, firms were required to get one permit to build a plant and a second to operate it.[14] This meant that a plant could be built but never allowed to operate. While this might sound far-fetched, it actually happened with the Shoreham nuclear power plant, which had received both its construction and operations permits but was closed due to public opposition before it ever operated commercially. Obviously, this created a serious risk for investors, and the regulatory reforms were an effort to mitigate that.

The growing opportunities in the nuclear industry were widely recognized then, as they are now. And while it's true that the subsidies in EPACT 2005 amplified enthusiasm for nuclear energy, the small number of reactors that would have received those subsidies couldn't have justified the broad investments that were taking place at the time. The fact is that there was authentic optimism about the industry, and investment followed.

In 2008, *U.S. News & World Report* called nuclear engineering the new hot job and reported that the industry also needed "skilled tradesmen and mechanical, electrical, chemical, and civil engineers with the know-how to run and build nuclear facilities."[15] Companies responded accordingly. For instance, one of the world leaders in nuclear energy, Areva (which now exists as two separate companies), added five hundred jobs at its headquarters in Lynchburg, Virginia, a 25 percent increase. The company also said it would add another nine hundred technical jobs in Wilmington, North Carolina. Those jobs would pay roughly $50,000 more than the average annual salary in New Hanover County in 2008.

At the same time, Westinghouse intended to continue expanding its labor supply, which had already been increased by thousands. The company, with its partner the Shaw Group, even announced plans to build the first commercial nuclear module fabrication and assembly facility in the United States. The idea of the moment was that new nuclear plants would be built in modules to save costs—another example of the industry innovating on its own. The Westinghouse facility was set to manufacture components for new and modified reactors and would have created thousands of jobs.

Even the federal government was preparing for a nuclear renaissance. The NRC had hired more than four hundred employees to handle new plant licensing and planned to expand its workforce even further to meet anticipated obligations. And of course all these new reactors would need to be fueled, so USEC (now Centrus), the only domestically owned enrichment company in the US, intended to build a new $3.5-billion plant in Piketon, Ohio. USEC estimated that this American Centrifuge Project would create more than three thousand jobs in Ohio and an additional three thousand direct and indirect jobs for its suppliers to manufacture the centrifuge parts.

Areva, meanwhile, selected Idaho Falls, Idaho, as the site of its $2-billion enrichment facility. It planned to begin operations in 2014 and to operate at full capacity by 2019. GE-Hitachi announced plans to build its Global Laser Enrichment facility in Wilmington, North Carolina, with construction beginning in 2009. Finally, Louisiana Energy Services (now Urenco) began building a $1.5-billion National Enrichment Facility in Eunice, New Mexico, to start operations in 2009 and reach full capacity by 2013. This was perhaps the most successful of these investments; it began operating in 2010 and has even undergone expansion.[16]

The nuclear industry didn't need the federal government to tell it where to invest. Without prompting from Washington, the Edison Welding Institute had put together a consortium of nuclear companies to identify supply chain weaknesses, prioritize objectives, and improve quality. The Nuclear Energy Institute had implemented a comprehensive program to achieve similar goals. These associations are how industry determines where capabilities must be strengthened.

Even the university sector was responding, with large universities, local community colleges, and everything in between expanding to meet the industry's demand for more engineers and skilled laborers. According to the Nuclear Engineering Enrollments and Degrees Survey, the number of science degrees granted by nuclear engineering programs had risen from 159 in 2000 to 346 in 2006, reflecting "substantial increases in enrollments reported in [those] years. The number of B.S. degrees in 2006 is the highest reported in the last ten years," it said at the time.[17]

It was no wonder that major universities were ramping up their nuclear engineering programs. The industry's high demand for engineers begets higher salaries, and higher salaries encourage greater enrollment in nuclear engineering programs. Purdue University had almost tripled the enrolment in its highly regarded nuclear engineering program since the mid-2000s. Texas A&M had one of the fastest-growing programs in the country, and the University of Florida had continued increasing enrolment as well as boosting its research grants. Other schools, such as the University of Virginia, were re-establishing their programs and expected to generate a great deal of interest. The upward trend in the number of nuclear engineering students was also generating a high demand for quality professors.

Meanwhile, community colleges were beginning to collaborate with private companies to offer training in skilled and craft labor. Duke Energy donated $1.25 million to North Carolina State University's College of Engineering, to help fund a professorship in engineering and advocate for the teaching of engineering in grade schools and high schools. Progress Energy, a utility company, awarded $60,000 to Florence-Darlington Technical College's Advanced Welding and Cutting Center to meet the projected increased demand for pipe welders, who have critical skills for nuclear plant construction. And the New Jersey–based Public Service Enterprise Group piloted an entry-level program at Mercer County Community College to provide training and education for specific technical jobs.

Yet nearly all of this investment came to a grinding halt as enthusiasm for new nuclear construction largely fizzled. For example, construction for the nuclear component manufacturing plant in Newport News, Virginia, actually broke ground before it was delayed due to the slow permitting process.[18] The project was eventually suspended indefinitely. A similar fate befell most of these projects.

WHERE DID THE GOOD TIMES GO?

EPACT 2005 was supposed to keep the good times rolling for the nuclear industry by encouraging ongoing investment. Instead, it did the opposite.

It failed for the same reasons that nearly all subsidies fail when their purpose is to prompt a successful commercial business. Subsidies undermine the two things that any successful commercial enterprise requires: a concrete grasp of the current market and an ability to prepare for a future that is essentially unknowable.

EPACT showed that subsidies alter how firms engage both the present and the future. The subsidies ignite enthusiasm that is driven to at least some degree by fleeting and unpredictable political preference rather than real-world market forces. Because of that, the success of any investments will depend on both political/taxpayer support and value to markets. Substantial shifts in either will undermine the strength of the investment.

This is because subsidies tend to erode pressure to continue tweaking the underlying economic and regulatory issues that inevitably emerge as new policies take effect. EPACT, for example, included production tax credits, finance subsidies, and insurance against regulatory delays. Instead of demanding that firms figure out how to produce power profitably and efficiently, EPACT incentivized them to take advantage of production tax credits to ensure profits, apply for taxpayer-backed loans, and rely on insurance to blunt the impact of regulatory delays.

In each case, firms were encouraged to seek public support to make their business models work, instead of doing the things that would result in a sustainable operation. This type of approach leaves individual firms and entire sectors dependent on government programs and less prepared to shift as markets change. And soon the markets did change in two big ways. First, natural gas prices began a steady decline with the innovation of fracking. Second, the regulatory efficiencies that were expected because of alterations to the design certification and permitting processes never really materialized.

Had efforts been focused on resolving underlying issues, the industry would have continued to grow informed by authentic market signals. It would have naturally uncovered and addressed ongoing structural policy and regulatory issues. Yes, it's possible that commercial nuclear still would not have been successful, but on the other hand, it may have.

An enterprise that exists purely in a free market will live or die based on its ability to meet market demand and is not required to divert attention and resources to the wants of government. Such businesses must remain agile, innovative, and responsive to inevitable market shifts. Computer gaming is a good example of this. The video game *Pong* was introduced for home use in 1975, when total industry revenue was around $15 billion.[19] Today, the industry is closing in on $200 billion in annual revenue.[20] (Some estimates double that amount.)[21] This growth was driven purely by market demand, and the most successful companies are those that were able to meet ever-changing gamer preferences while simultaneously preparing themselves to compete in the future, even without knowing what that future would hold. This approach rewards innovation and competitive pricing.

The other extreme is an enterprise that depends almost exclusively on a government customer. That business will live or die based on sustained government support and will exist largely outside of commercial market demand. Firms that make combat equipment like fighter planes and battle tanks are good examples of this. While these firms often present themselves as proponents of innovation and competitive pricing, the reality is that their success is not necessarily defined by either. Instead, it depends on getting enough politicians to support purchasing their products. That requires not just offering a good product at a competitive price but also offering products that meet a litany of political preferences, such as where they are manufactured, where they will be serviced, who will use them, and so on and so forth.

A business that made fighter jets for government and private customers would probably fail because the market for private fighter jets is presumably zero. If the firm's success depended on both private and public sales, it would be highly incentivized to seek additional taxpayer support to make up for the lack of profits from private sales. When a company is built to depend on both, it is less able to withstand disruptions of either, especially when the product being sold is meant to be commercially viable. Authentic market demand may exist, but subsidies create unwarranted enthusiasm. This in turn incentivizes overinvestment. For example, a video gaming company that expands capacity to meet some anticipated government purchase or to take advantage of some government subsidy will have overinvested if the taxpayer money to buy those games is never appropriated or if the market meant to be served by the subsidized investment never materializes.

This is what happens in the energy industry when government intervenes. It creates expectations and encourages investments that are informed more by politics than by market fundamentals. At least to some extent, that is what happened with EPACT and nuclear energy. When the market shifted, the industry was ill prepared to adjust, and the underlying policy and regulatory issues remained in place.

DÉJÀ VU ALL OVER AGAIN

Fast-forward to today, and the same exact thing is happening.

Interest in expanding nuclear power began growing a few years ago as energy producers came to understand that they would be unable to meet the growing power demands of a modern economy under the strict regulatory regime being imposed to reduce carbon dioxide emissions. Eventually, policymakers realized that intermittent energy like wind and solar alone would be inadequate to meet CO_2-reduction goals and nuclear would be the only feasible way achieve them. This created something close to authentic enthusiasm for nuclear energy. It is not genuinely authentic, though, because the market to reduce CO_2 is not real or sustainable in its current form.

The government will never be able to reduce CO_2 with a top-down, imposed transition away from hydrocarbons. It will be too costly and have virtually no perceptible environmental benefits. According to an analysis by my colleague Kevin Dayaratna, "Under the most modest estimates, just one part of this new deal [the Green New Deal] costs an average family $165,000 and wipes out 5.2 million jobs with negligible climate benefit."[22] But the enthusiasm is authentic in that industry proponents and investors see the potential that nuclear has for future power generation.

Some enthusiasm may also be about CO_2 reductions, but it is primarily driven by nuclear energy's ability to provide safe, clean, affordable power in massive quantities. Indeed, you could reasonably conclude that the real impact of the political push to reduce CO_2 is that it has provided politicians who might have actively sought to stop nuclear in the past a reason to at least accept it—if not overtly support it—moving forward. Further, it provides investors with some semblance of hope that government will not once again be used to stop nuclear energy in its tracks. Indeed, the push to reduce CO_2 could be the catalyst to force additional reform.

The American economic system depends on the premise that market competition yields better results over time than a managed economy. The current approach to energy policy, where targeted subsidies and preferential treatment from the government essentially encourage firms to focus more on influencing politics to protect their narrow interests, contradicts what we know works economically. That's because the approach breeds cronyism, stifles innovation, and disincentivizes companies to be competitive with other power-generating sources.

Subjecting the nuclear industry to the free market will not guarantee that all of the nation's nuclear firms will be successful, or that the wider industry will look in the future how it has looked for the past five decades, but it does promise that whatever industry emerges will be innovative and provide Americans with the energy they want at a price they are willing to pay. That's why lawmakers should strive to remove policy distortions that corrupt the underlying structure of competition across all energy sectors.

In 2018, my former colleagues Katie Tubb and Nick Loris described aptly how American families and businesses benefit from markets.[23] They wrote:

> **Markets are customer-centric.** Competition in electricity services allows greater customer choice through the power of the consumers' own dollars rather than through the disconnected votes of a small panel of public utility commissioners. Consumer choice comes not only in the form of resource choice (renewables, conventional fuels, or a mix) but also in financial choices (for example, fixed rates, risk preferences, indexed rates, or short-term or long-term contracts). In the end, because electricity providers have to work for their customers, prices are competitive and quality improves.

Markets enable organic innovation. Technology and energy-source neutral competition in electricity markets allows the endless creativity of people to meet customer energy needs and preferences *while* protecting customers from unwise investments. In contrast, regulated monopoly power regions are a "fundamentally permission-based system" where investments require highly political negotiations and approval from commissioners, or are mandated from legislatures chasing the latest technology flavor of the day.

Political interventions destroy investment confidence in the face of ever-changing and arbitrary political winds. Further, while "top-down, integrated resource planning approaches are tempting because it is easy to think that experts know exactly the right mix and location of generation resources," experts are often wrong or slow to change and [consumers] have to foot the bill.

Markets efficiently align incentives. Technology-neutral competitive markets allow prices to communicate accurate information to producers and consumers about the value and cost of electricity generation and delivery. Consequently, competitive markets force power suppliers and investors to consider the costs and benefits to their customers, and incentivize the discipline to be more efficient—in operations, investments, and regulatory compliance—than competitors. Monopoly structures on the other hand guarantee that some, if not all, costs of service are negotiated by utilities and public utility commissions, incentivizing [energy firms] to increase spending within the margin of what is politically feasible so as to increase profits.

Ultimately, markets should determine the fate of the nuclear industry. That doesn't mean that policymakers have nothing to do. They absolutely do, beginning with unwinding the decades of economic distortions that they've foisted onto the energy industry broadly and overhauling the onerous regulatory burden that encumbers the nuclear industry specifically.

But that's not what is happening. In fact, policymakers are doing the exact opposite. Rather than allow the nuclear industry to evolve organically, Washington has once again stepped in. Only this time, it is doing so on a massive scale with the Inflation Reduction Act and other recent legislative efforts. This is in addition to the bevy of nuclear spending that is essentially baked into the baseline of Department of Energy funding.

There is no question that efforts to revive the nuclear industry with subsidies in the mid-2000s were a failure. Yet policymakers and many supporters of nuclear energy seem bent on carrying out the same approach and in a heightened fashion, with even more subsidies and more government intervention.

To get a different outcome, the United States needs a new approach.

CHAPTER 8
THE PROMISE, CHALLENGE, AND OPPORTUNITY OF ADVANCED REACTORS

Could a big future for nuclear energy be driven by small reactors? Could they avoid the financial and regulatory barriers that we've discussed? They could expand nuclear energy's market. But they could also fall victim to the same government excesses that have hindered traditional nuclear. Let's take a closer look.

SMRs share many of the attractive qualities of large reactors, such as the ability to provide abundant emissions-free power, but they may be faster and less costly to build, and they can be used in more remote locations or to power discrete facilities like industrial parks or data centers. Advanced reactors may not be ready to take the place of traditional large reactors, but they represent an important potential growth area for the commercial nuclear industry. In fact, if their promise is realized, the industry would be transformed.

Large reactors are very expensive to license and construct, and they require massive up-front capital investments. Recent experience in the United States, France, and Finland tells us that new large reactors can cost well over $10 billion and must be financed all at once. Smaller reactors, while providing far less power, can be built in modules, which allows them to be added to and paid for over years or decades as power demand grows.

That some recent efforts to build SMRs have so far met with financial and technical challenges doesn't mean that past experience defines future prospects. Of course, if we continue to rely on the same regulatory and policy regime, then we should expect the same outcome. But if that regime is reformed, we can reasonably assume that the cost for SMRs will come down. What's more, a less subsidy-dependent industry will be incentivized to find solutions to whatever technical or financial challenges arise.

Advanced reactors come in all shapes and sizes, and this makes them ideal for an array of different applications. For example, they could be used for small or isolated communities that don't need to generate a lot of power, for parts of the world that lack energy infrastructure, or for energy-intensive industries that want to supply their own power. The potential applications are truly endless. It's still early days, but many utilities are already voicing a commitment to SMRs. Some have even announced partnerships with SMR companies to advance projects, and a 2022 survey by a leading industry group found that utilities are currently interested in building up to three hundred SMRs in coming years.[1] Though amplified, this interest is nothing new. In 2005, the Alaskan town of Galena announced that it was in talks with Toshiba to set up a tiny reactor that would run almost unattended.[2] Some power-intensive industrial giants have also voiced past interest in pursuing SMRs.

Advanced reactors are critical to the future success of nuclear energy because of all they potentially have to offer. They could cut electricity costs in isolated areas because they would take the place of expensive transmission lines. Or some technologies could perhaps be integrated into existing energy infrastructure—by being built into old coal plants, for example, which is one reason both Virginia and West Virginia are so interested in the technology.[3] The reactors would replace the coal boilers and could potentially be hooked into the existing turbines and distribution lines—which is not to say that we should shut down perfectly good coal plants to make way for SMRs!

Advanced reactors can also be used in a variety of applications that have substantial power and heat requirements. The chemical and plastics industries and oil refineries, for instance, all use massive amounts of natural gas to fuel their operations. And small reactors could produce the heat needed to extract usable oil from oil sands, which also currently require large amounts of natural gas. Then there is the tech industry, which increasingly needs reliable electricity to power everything from essential infrastructure like data centers to activities like crypto mining.

Natural gas prices vary significantly over time, making the long-term predictable pricing of nuclear very attractive. At the same time, large electricity grids are becoming less reliable as policymakers force renewables onto networks that were never designed for intermittent energy sources. And advanced reactors may provide a practical solution for desalination plants (which require large amounts of electricity) that can bring fresh water to parts of the world where supplies are depleting. Perhaps most important, advanced reactors have the potential to bring power and electricity to the 675 million people who don't have it and the 2.3 billion who rely on biomass (wood, agricultural residue, and dung) for cooking.[4]

Small reactors will add a new dimension to reactor competition as multiple designs emerge on the market. This increased competition will not only create a robust market but also provide an additional incentive for large reactors to improve. If smaller reactors begin to capture a share of the nuclear market and the energy market at large, that will drive innovation and ultimately lower prices for both new and existing technologies.

Although the nuclear industry shrank over the past decades, much of the domestic infrastructure remains in place and could support the expansion of small-reactor technologies. Many of the skilled workers required—welders, electricians, steamfitters, etc.—are already out there, working in other industries, and some companies are partnering with colleges to train more. A number of companies are also expanding their manufacturing, engineering, and uranium enrichment capabilities—all in the United States.

These are all the reasons that advanced reactors have emerged once again as a central piece of the nuclear conversation. Companies of all sizes are investing in these smaller, safer, and potentially more cost-efficient reactors. Around the world today, there are approximately 22 GW of advanced reactor projects, valued up to $176 billion in investment, in some stage of development.[5] The total value of the global advanced reactor market could reach $300 billion by 2040.[6] That is a ton of potential and opportunity. The US is currently leading the world, with over 18 percent of that investment.[7] China and Russia are also emerging as major players. Utilities and other energy-intensive industries are even forming partnerships with universities, laboratories, and reactor designers to prepare for potential future construction.[8]

IS PAST PROLOGUE?

There is a lot to be excited about, but much could still go wrong. As we've seen over the past couple of decades, planned nuclear projects do not always (or nearly ever) result in completed power plants. Despite market-driven enthusiasm prompting early private interest, which is then bolstered by some package of government subsidies, it seems that commercial nuclear markets rarely live up to early expectations.

In fact, that is precisely what happened with advanced reactors in the past. Even though they're often presented as an alternative new technology, small reactors have actually been around for a long time. The feasibility of smaller reactors was part of a process the US government initiated with industry partners in the early 1950s to determine which reactors made the most commercial sense. Interestingly, the Atomic Energy Commission required industry groups that participated in these early studies to do so at their own expense.[9] Even back then, government officials understood that while federally funded research was sometimes appropriate, only the private sector could drive commercialization.

At any rate, this process generated substantial interest in an array of commercial nuclear technologies. Of note (and relevant to the current debate over nuclear energy), government financing was not needed to build some of the earliest plants in the late 1950s and 1960s. In fact, some were built entirely through private finance, while others were chiefly financed privately with some research support from the government. In total, Washington commissioned seventeen smaller reactors in those early years.

One of the most successful efforts, the Dresden 1 boiling water reactor outside Chicago, which operated from 1960 to 1978, was privately financed through a cost-sharing agreement among eight companies.[10]

The Peach Bottom high-temperature gas-cooled (HTGC) reactor, which operated commercially from 1967 from 1974 near Philadelphia, got public funds for some preconstruction research and development but was primarily financed through a consortium of over fifty utilities. Commercial light-water reactors in the US are either boiling water reactors (BWRs) or pressurized water reactors (PWRs).[11] BWRs generate steam directly from water heated by fission in the reactor core to drive a turbine generator. PWRs use high-pressure water heated in the core to heat a separate water source, creating steam to power the turbine. Interestingly, while none operate commercially today, HTGC reactors have been commercialized. These reactors use helium gas as a coolant, which is heated to very high temperatures by fission in a graphite-moderated core and can then be used to generate electricity through a gas turbine.

The point is that a young, robust, innovative, and growing advanced reactor industry was emerging in the United States from the very dawn of America's commercial nuclear era, and interest in it was driven by market potential, not subsidies. That's not to say that government did not have a role—it obviously did. But policymakers at the time took great pains to focus on ways to move the technology out of the lab and into the market without supplanting the critical role of private investment and finance. That certainly doesn't mean that every investment was a winner, but it does demonstrate an early recognition that markets would be key in driving commercial nuclear forward.

Yet despite some early successes, America's nascent SMR industry largely faded, giving way to the large light-water reactors that make up the country's commercial nuclear fleet today. The irony is that the preference for LWRs had almost nothing to do with what technology made the most commercial sense and almost everything to do with a decision from government.

As it turns out, America's push for reliable naval propulsion combined with a desire to quickly commercialize any nuclear technology led government officials to focus on light-water technology, which was considered a more stable and mature technology, and thus more appropriate for naval propulsion. The entire program for LWRs was measured on reliability and performance standards as they related to the navy's needs and preferences, not their efficacy as a commercial offering. The result of that decision was that the entire government apparatus to allow nuclear technology to be commercialized revolved around light-water reactors. Government led, and industry followed.

Perhaps it is too dramatic to state that at that point, the die was cast on the economic viability of nuclear power and the role of advanced reactors. But it was without question a public policy decision that set commercial nuclear power on a path that subordinated commercial viability to the government's desire to prioritize certain reactor types over others. And this led the nation to where it is today, for better or worse.

Despite this, the United States dipped its toes into the SMR pool once again around 2010. Rising prices for large reactors prompted a spike in SMR interest, and private investors and entrepreneurs began dedicating substantial resources to the technology.[12] Arguably, this interest was based on authentic market potential, as enthusiasm around nuclear energy remained high but the market for large nuclear reactors began to soften. At the time, there were virtually no government mandates for carbon-free energy or SMR subsidies, and this mini renaissance was occurring despite a regulatory bias in favor of large light-water reactors.

Enthusiasm around small nuclear grew organically because SMRs provide flexibility that could solve a variety of energy challenges. Many communities and organizations explored the possibility of using SMRs, including towns in Alaska, which saw the potential for small nuclear to provide affordable energy to remote places. In Canada, Western Troy Capital Resources planned to form a private corporation to provide electric power from small reactors to isolated locations. Grays Harbor, Washington, also showed interest, with public utility officials discussing the possibility of installing multiple small nuclear plants with NuScale Power Corporation.

During this period, several companies were also developing SMR designs. Toshiba was developing their 4S mini reactor and Babcock & Wilcox (now BWXT) introduced their mPower reactor, a 195 MWe integral pressurized water reactor, in 2009. Westinghouse was developing a 225 MWe SMR design, while NuScale Power was working on their 45 MWe PWR design and General Atomics was developing their Energy

Multiplier Module (EM2) reactor. Many of these efforts have been paused, permanently shelved, or reimagined as new projects. Thus far, only the NuScale SMR was certified by the NRC (in 2023),[13] while General Atomics is still undergoing pre-application activities as of this writing.[14]

But policymakers were unable to leave well enough alone. Just as private enthusiasm for large reactors had led to EPACT 2005 subsidies, enthusiasm for SMRs prompted Congress and the Department of Energy to offer a host of new programs—underwriting everything from licensing fees and R & D to construction costs—in an effort to move advanced reactors forward.[15]

The results were virtually the same: millions, if not billions, of public dollars were spent, and no new reactors were built.

There is obviously a cycle here. Commercial interest in advanced reactors emerges because of their potential. This invites taxpayer and policy support to help rapidly commercialize the technology. Despite this help, or maybe because of it, the technology runs into regulatory, cost, and technical barriers. Few, if any, reactors get built, and eventually interest fades away.

Why does this keep happening?

RICKOVER'S PARADIGM

Admiral Hyman G. Rickover, the Father of the Nuclear Navy and an outspoken advocate of civilian-use nuclear power, went a long way toward answering that question in his 1953 "Paper Reactor" memo.[16] Admiral Rickover obviously had an interest in advocating for light-water technology, and you could argue that his memo was more about diminishing the prospects of technological competitors than shedding light on the challenges of introducing new nuclear technologies into the market, but his thoughts on the matter are still relevant. In the memo, Rickover describes his perspective on the difference between what he termed "academic" and "practical" reactors. An academic reactor, Rickover explained, is characterized by the following:

1. It is simple.
2. It is small.
3. It is cheap.
4. It is light.
5. It can be built very quickly.
6. It is very flexible in purpose ("omnibus reactor").
7. Very little development is required. It will use mostly off-the-shelf components.
8. The reactor is in the study phase. It is not being built now.

A practical reactor, on the other hand, has the following characteristics:

1. It is being built now.
2. It is behind schedule.
3. It is requiring an immense amount of development on apparently trivial items. Corrosion, in particular, is a problem.
4. It is very expensive.
5. It takes a long time to build because of the engineering development problems.
6. It is heavy.
7. It is complicated.

As history has demonstrated time and again, Admiral Rickover's insightful analysis aligns nicely with America's commercial advanced reactor experience. What is presented as simple, inexpensive, and close to market-ready always seems in reality to be complicated and expensive and never quite market-ready. Today this seems to be true for reactors big and small.

As we contemplate the potential of advanced reactors moving forward, we would be wise to remember Admiral Rickover's words. In a few short sentences, he captures what has too often been the experience of making an "academic" reactor into a "practical" one, and perhaps that explains what has gone wrong in the past.

The obvious question, then, Is how do we move beyond Rickover's paradigm?

Policymakers were on the right track when they first began trying to move nuclear technology from the labs into the market in the 1950s. They relied on the private sector to identify which technologies made the most sense for commercial purposes and then leaned on those with a commercial interest in nuclear power to foot the bill. This approach, had it been carried forward, would have properly aligned incentives and responsibilities. Those early officials understood that nuclear power had no commercial future if it wasn't profitable, and that only the private sector could move it in that direction. They also understood, however, that nuclear energy was a new technology that existed only in government labs, and that government needed to create avenues through policy for the technology to migrate into markets.

Unfortunately, the decision was made—largely by Rickover, as it turned out—to focus government efforts on a specific technology and to prioritize other considerations over economics. Even though some later attempts were made to bring other technologies forward, this decision began the process of using policy to ensure that America's commercial nuclear might was defined by a single technology.

From a national standpoint, this decision was arguably the right one. It led to a world-class nuclear industry, provided the industrial base to support the growing nuclear navy (which helped win the Cold War and remains extraordinarily successful today), and helped to define a generation of peaceful nuclear technology that remains the global standard.[17] But what it didn't result in was an economically sustainable, competitive, innovative, and independent nuclear industry. That decision to subjugate economics to government preference created a major distortion that continues to reverberate today.

Currently, American policymakers are tripling down on this approach. Whether it's through the Energy Policy Act of 2005, the Bipartisan Infrastructure Bill, or the Inflation Reduction Act, lawmakers have shifted into higher gear in their efforts to use taxpayer money to drive the advanced reactor industry forward. Hundreds of billions in taxpayer-backed loans are now available to fund nuclear projects, $2.5 billion is available to fund demonstration plants, and a plethora of tax subsidies have been authorized.

The problem with this approach is that it ignores the larger systemic problems that create the unstable marketplace to begin with. These systemic problems generally fall into three categories: licensing, nuclear waste management, and government intervention. When it comes to licensing, the NRC is not optimally prepared to efficiently process applications to license a technologically diverse fleet of new reactors, but no reactor can be offered commercially without an NRC license. This problem has been recognized for some time. In a September 2009 interview, former NRC chairman Dale E. Klein said that small nuclear reactors pose a dilemma because the commission is uneasy with new and unproven technologies and feels more comfortable with large light-water reactors, which have a long safety record.

Thankfully, this reticence is beginning to change, and the NRC is increasing its engagement on advanced reactors. In fact, it has issued a certification for one SMR design (albeit a downsized version of a light-water reactor), which means that design can be used in commercial applications. Currently, there are seven developers engaging the NRC to possibly submit applications to build plants using advanced reactor designs, and one company has applied to construct and operate a new plant.[18]

There is also a major effort underway within the NRC to develop a technology-neutral, risk-informed, performance-based regulatory framework for advanced reactors as set out in the Nuclear Energy Innovation and Modernization Act (NEIMA), which Congress passed and President Trump signed in 2019.[19] NEIMA not only directs the NRC to develop an updated regulatory framework for new reactor designs but also requires it to ensure that it has adequately trained staff to implement the new process. Though criticism of the rule's initial draft suggested that the new framework remained overly burdensome and complex, it seems that that the NRC has taken those critiques and is working to improve it.[20] The commission is expected to complete its process by 2027. While this is certainly a step in the right direction, we are correct to remain skeptical of any effort to modernize existing institutions to allow nuclear technology to reach its potential, and policymakers should continue seeking alternatives that would allow nuclear to move forward through other processes.

Then there is the issue of nuclear waste. The lack of a sustainable solution to nuclear waste management is perhaps the greatest obstacle to a broad expansion of US nuclear power. It is especially detrimental to advanced reactor commercialization because SMRs could be key in solving the nuclear waste issue. Because advanced reactors consume fuel and produce waste differently than LWRs (some are even designed to *consume* nuclear waste), they could contribute greatly to an economically efficient and sustainable nuclear waste management strategy. But without a properly functioning market-driven program, the attractiveness of advanced reactors in helping to deal with nuclear waste is essentially erased. This subject will be investigated further in a later chapter.

The federal government has failed to meet its obligations under the 1982 Nuclear Waste Policy Act, as amended, to begin collecting nuclear waste for disposal in Yucca Mountain. The Obama administration's success in shuttering the Yucca Mountain spent fuel repository program without having a backup plan worsened an already difficult situation. This outcome was predictable, though, because the current program, as defined by the Nuclear Waste Policy Act, is based on the flawed premise that the federal government is the appropriate entity to manage nuclear waste. Under the current system, waste producers can largely ignore long-term waste management because they are not responsible for it.

The key to a sustainable waste management policy is to connect financial responsibility for waste management to waste production. This will increase demand for waste-efficient reactors and drive innovation on waste-management technologies, such as reprocessing or the use of advanced reactors.

AN OPPORTUNITY TO BE LOST OR GAINED

Getting policy right is important to the United States for several reasons. First, advanced reactors could provide needed power to ever-evolving energy markets. It would be best if Washington, especially the Department of Energy, just got out of the way and allowed energy producers to conduct business in the ways they deem most appropriate, but the fact is that Washington's preferences will continue to limit, at least in part, what energy sources are available to Americans.

It's equally important to take advantage of the opportunity that emerging demand for advanced reactors could represent. As we've seen, the domestic and global markets for advanced reactors are potentially enormous, with current predictions measured in the hundreds of billions of dollars. And the market's potential could be even greater given the large swaths of the world's population that have too little access to electricity or no access at all. That is why advanced reactors are gaining momentum in many countries. In fact, Russia and China are arguably best positioned to lead the world in meeting this growing demand.

Russia in particular has a long history of successfully developing and exporting nuclear reactors. Of the approximately sixty reactors currently being built around the world, Russia is constructing twenty of those in seven different countries.[21] As mentioned earlier, Russia also enjoys commercially viable nuclear fuel and waste management sectors, which allows it to offer full fuel cycle commercial nuclear options on the export market. Combined, this has enabled Russia to establish a clear lead when it comes to commercial nuclear exports.

Now Russia is translating its expertise in developing SMR technology into domestic energy production and exports. Construction has begun on its first land-based SMR, which will provide power to the remote northeastern Republic of Sakha, and plans are in the works to build more.[22] To put action behind its goal to supply up to 20 percent of the emerging global market for modular reactors, Russia recently signed a contract with Uzbekistan to export six SMRs. This is the world's first such contract. Unlike other agreements that have been inked in recent years, this one is for actual construction, which is set to start as early as 2024.[23]

No one can predict what future nuclear demand will look like, but if an SMR market does materialize, there is little question that Russia, with its expertise in commercial nuclear power and its demonstrated commitment to nuclear energy exports, will continue to maintain a top spot.

Then there is China. While it has yet to become a major nuclear exporter, it is developing world-class expertise in the commercial nuclear industry with its massive domestic construction program. It has 55 reactors currently in operation, with another 26 under construction, 41 planned, and an astounding 158 in the proposal stages.[24] Interestingly, it took China only around ten years to build the amount of nuclear capacity that the United States needed four decades to create, and it's building efficiently and cost-effectively.[25] One of the reactors under construction is an SMR.

China is now taking what it has learned by building reactors domestically and beginning to compete in the global market. Through its Belt and Road Initiative, it aims to acquire up to 30 percent of the global commercial nuclear market.[26] This should be considered a floor and not a ceiling. While Russia is currently the clear leader in global exports, its invasion of Ukraine could create substantial downward pressure on demand for its reactors. This would open export opportunities for other nations, and China certainly seems capable of successfully competing for that opportunity.

Though China and Russia are well positioned to dominate the future of commercial nuclear power, so too is the United States. Despite their recent successes, neither country can outdo American technology, innovation, or entrepreneurial spirit. And the US does not need to face its competitors alone. It has relationships with countries like South Korea, which, like Russia and China, has been successful in recent years in building reactors domestically and exporting them elsewhere.

Ultimately, an industry built on free enterprise and private-sector ingenuity will always outcompete one managed by politicians and bureaucrats. The only question is whether policymakers will recognize this opportunity and allow a uniquely American industry—one that is competitive, innovative, and driven by free enterprise—to emerge.

CHAPTER 9
FUELING THE TWENTY-FIRST CENTURY

The United States gets around 19 percent of its electricity from ninety-four commercial nuclear reactors, and these reactors are powered by uranium fuel (although future reactors could be fueled by other sources). *The good news is that to produce the same amount of power as other energy sources, like coal, natural gas, and petroleum, nuclear energy requires far less fuel. The bad news is that the United States imports 95 percent of its uranium.[1] About a third of that comes from friendly countries like Canada, Australia, and Namibia, but until recently, we also depended on Russia for around 12 percent of that uranium.[2] After the invasion of Ukraine, the government banned the import of Russian uranium, but the Department of Energy has waived the ban for some imports—more on that later. How much Russian uranium is ultimately brought in is unclear at the time of writing.

Okay, so we rely largely on friendly nations for our nuclear fuel instead of producing our own. That doesn't sound too bad. The problem, though, is that freshly mined uranium can't power American commercial reactors. For natural uranium to be useful as fuel for US reactors, it must first go

---

*I must acknowledge CJ Milmoe's contribution to this chapter. CJ is a longtime nuclear industry expert and a consultant on nuclear energy development and trade regulation. His work in these important areas has informed my thinking on the issues discussed here.

through a process called enrichment—and this is where our dependence on Russia became difficult.

Uranium is a readily available mineral around the world, but the ability to enrich that uranium for use in nuclear power plants is far more limited, and Russia controls around 46 percent of that market.[3] The United States, by contrast, controls only around 8 percent of enrichment capacity, even though it is by far the leader in nuclear energy generation.[4] In fact, the US enriches only around 20 percent of its own uranium. Some of its imports are provided by countries like the United Kingdom, Germany, and the Netherlands, but traditionally about 25 percent came from Russia.[5]

Interestingly, this was not always the case. Though there were some ups and downs, the United States produced much of its own uranium until 1980, when production began to decline and never recovered.[6] And the trend was the same for enrichment.[7] The US was largely self-sufficient in uranium production and enrichment until the end of the Cold War and was even experimenting with alternative nuclear fuels in earlier decades, but those days are long gone.

Today, the nuclear fuel industry, predominantly uranium-based, could be about to undergo significant changes once again. First, and most obviously, building a new generation of reactors would likely put substantial pressure on current uranium and enrichment supplies. Second, and more acutely, there is a pressing need for peaceful nations to stop using Russian nuclear fuel.

Uranium producers have begun responding. Uranium production from mines surpassed 49,000 tonnes in 2022,[8] and domestic uranium mining activities have increased substantially in recent years. In 2023, the number of exploration and development drilling holes increased from 1,008 to 1,930.[9] This is for good reason. According to the World Nuclear Association, uranium requirements for fueling reactors could increase to nearly 63,000 tonnes by 2030, and closer to 100,000 tonnes in the decade after that.[10] Given that more than half of the world's uranium production comes from just three countries (Kazakhstan, Canada, and Namibia), the US has many incentives to increase its domestic uranium mining.

Markets are reflecting these anticipated shifts in uranium fuel markets. Indeed, uranium prices, which have been on the rise for some time, have recently hit a fifteen-year high.[11]

GLOBAL PRICE OF URANIUM

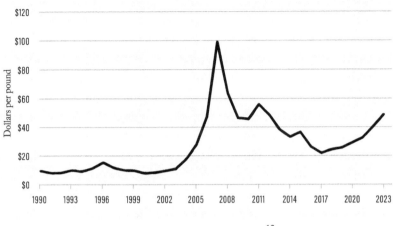

SOURCE: Trading Economics.[12]

This is one reason why the United States and its friends and allies must consider tapping more of their own uranium reserves and expanding their enrichment capacity. The US should also consider using alternative fuel approaches, like thorium or the reuse of spent nuclear fuel. Unfortunately, burdensome regulations, politics, and bad policy once again hamper access to available energy resources across the country. The nation can now add uranium to the list of resources that local, state, and federal bureaucrats are increasingly deeming off-limits.

For example, the nation's largest known uranium deposit was discovered in the 1980s on a farm in southern Virginia. The owner of that land has explored the possibility of mining the approximately 130 million pounds of uranium (worth over $6 billion at today's prices) believed to be on the site.[13] But even though uranium has been mined safely around the world for decades, including in New Mexico, Nebraska, Utah, and Wyoming, Virginia bureaucrats, like those in other parts of the country, have decided to prohibit the owner from developing that resource.

More recently, the Biden administration made a major tract of federal land off-limits to uranium development when it established the Baaj Nwaavjo I'tah Kukveni—Ancestral Footprints of the Grand Canyon National Monument.[14] The monument designation permanently banned any new mining claims, including for uranium, on the nearly 1 million acres covered.[15]

Some may think that such a policy is reasonable: no one wants to see an industrial mining operation in the Grand Canyon. And respect for ancestral lands is laudable. But that's not the real story. The Biden administration presented the decision as a choice between protecting the Grand Canyon and the surrounding ancestral lands or allowing mining to occur (which, by implication, would destroy both). This is the same sort of false framing that has consistently been used to ban energy development in the United States, especially around nuclear-related activities.

The truth is that the modern mining industry is perfectly capable of undertaking commercial operations while also protecting public health, safety, and the surrounding environment.[16] Indeed, strict state and federal regulatory oversight demands that mining operations are safely carried out, and that disturbed landscapes are appropriately restored.[17]

What's more, the designated area lies to the north and south of the 1.2 million acres that make up Grand Canyon National Park, so there is no threat that a tour of the canyon would one day be capped off with a pit stop at a uranium mine. Unfortunately, this is just one more in a long line of Biden policy decisions that create barriers to domestic energy development and mineral mining.[18]

While Biden likes to talk about green energy and energy independence, his policies toward reliable alternatives like nuclear and natural gas make it almost impossible for the American mining industry to develop the resources necessary to manufacture and fuel the president's vision. And in this case, it matters a lot given the potential increase in demand for uranium fuel and the meager infrastructure available to support it.

The problem is that opening new mines in the United States is extremely difficult, and federal and state-level decisions to take domestic supplies out of service make responding to foreign supply disruptions more challenging for the domestic uranium fuel industry.[19] Not only are domestic uranium miners missing out on the opportunity to provide secure supplies of uranium to American reactors, but reactor operators also have no choice but to continue their dependence on foreign suppliers.

The president's supporters say his monument designation and similar actions protect us and our environment from uranium mining, but no one wants to mine in the Grand Canyon or anyplace similar. To suggest as much is disingenuous. The reality is that these policies actually protect foreign uranium suppliers from US competition and prevent American reactors from accessing domestic fuel supplies.

THE URANIUM SUPPLY CHAIN

Many people think of the nuclear industry as only the reactors that are owned and operated by companies in the electric power industry. But it's much broader than that. A vital supply chain provides the fuel that enables those reactors to produce power, and that supply chain includes several steps:

1. **Mining and milling.** Natural uranium ore is mined and refined into a uranium oxide powder called yellowcake. Until the 1970s, 100 percent of US yellowcake came from domestic sources. A combination of misguided government policy and changing markets ended that. By the 1990s, US uranium mine production had shrunk to almost zero.[20] Meanwhile, opposition to uranium mining and heavy-handed regulation have increased the cost of domestic production to the point that reactor owners have turned to lower-cost foreign suppliers. Today, US reactor owners get only about 5 percent of their milled uranium domestically. Until recently, Russia supplied 12 percent of America's yellowcake, and its ally Kazakhstan supplies 35 percent. Rosatom, the state-owned Russian nuclear company, controls mining operations in both these countries. However, the global uranium mining base is broad and diverse, and friendly countries like Canada, Australia, and Niger are major suppliers.[21]
2. **Conversion to gas.** To be useful as fuel in most reactors, yellowcake uranium must be converted to uranium hexafluoride

gas for enrichment. In the early days of the US nuclear industry, all conversion was performed domestically. But a uranium conversion plant in Metropolis, Illinois—the only US plant—was idled in 2017, when low demand and low-cost competition made continued operation uneconomical.[22] Importantly, that plant recently reopened and already has orders filling capacity through 2028—another signal that the market is working.[23]

3. **Enrichment.** This is the key step in the fuel supply chain. Natural uranium contains only about 0.7 percent of the fissionable uranium 235 isotope (U-235) needed for reactor fuel. The converted natural uranium gas must be enriched to increase the percentage of U-235 up to 3 to 5 percent. This is called low-enriched uranium, or LEU. The cost and price of enrichment is driven by the amount of energy required to produce LEU, which is measured in separative work units, or SWU. This is where reliance on Rosatom is a problem for American reactor owners. The US does not have enough enrichment capacity to meet domestic demand. There is currently just one operating enrichment plant in the US—Urenco in New Mexico—and it has a capacity of only 4.7 million SWU per year.[24] US reactor owners require about 14 million SWU per year.[25] In 2021, we imported about 11.5 million SWU, with roughly 4 million SWU coming from Rosatom, the world's dominant supplier.[26] But here, too, natural market signals are driving expansion, with Urenco announcing plans to grow its operation in the United States.[27]

4. **Fabrication of enriched uranium into solid pellets and reactor fuel rods.** Once the uranium is mined, milled, and enriched, the LEU is turned into fuel pellets, which are then placed into rods and combined into bundles that fuel reactors. The US has adequate fabrication to serve domestic and export customers. Additional fabricators exist in friendly counties like France, South Korea, and Japan.

The issue with Russia goes beyond uranium. Prior to Putin's invasion of Ukraine, the US was importing a great deal of Russian energy, including oil and other petroleum products, but not enough to be considered dependent (in the sense that Russian energy imports could be replaced). Uranium is a different story, though, because the US imports substantial amounts of ore and even higher amounts of enriched uranium. It is arguable that these imports, especially on the enrichment side, reduced America's flexibility in how it chose to respond to the Russian invasion. For example, while the United States called on other countries to end their commercial relationships with Russia, it continued to import Russian uranium.

This is one reason why President Biden's push to ban Russian gas and oil imports soon after the invasion enjoyed broad support, but a ban of Russian uranium fuel was slower to gain traction. This was unfortunate because even though the US is more reliant on Russian uranium than Russian oil, those markets have many similarities, and obviously, the underlying moral and strategic justification for both bans is the same. Of course, banning Russian uranium does come with risks and will have costs, but like petroleum, uranium and related fuel services are global commodities, and those markets will adjust.

Still, given the US reliance on Russian uranium and the fact that no one at that point could predict how the Ukraine invasion would unfold, it was perhaps justifiable to avoid a ban at that early date. It is also the case that some portion of enrichment services were being purchased under long-term contracts at very low prices relative to the current spot market. By abrogating those contracts, the US would have effectively made it possible for Putin to receive more money by offering those services to the lucrative international spot market than he would have received by continuing to fill the contracts with US customers.

Nonetheless, it soon became clear that something had to be done. Rosatom was providing vital support for President Putin's military objectives, including the invasion, and it had taken over operation of Ukraine's Zaporizhzhia nuclear plant and threatened plant workers and their families.[28] The company also had provided economic leverage over countries dependent on the company for nuclear fuel and related services like waste disposal. It was obvious that for reasons both political and economic, the United States had to end its reliance on Russian nuclear fuel.

A PROBLEM CREATED IN WASHINGTON

With Russia's invasion of Ukraine, Putin returned to his established playbook of using energy as a weapon. Whether he wanted to cut off supplies or not, the West could hardly continue the purchase of Russian energy (or much anything else) in the face of this aggression. This was particularly disruptive given Russia's place as one of the world's top energy producers.

Even before the invasion of Ukraine, there was risk involved in relying on the Russian uranium fuel supply, but that risk was correctly the responsibility of private industry to understand and mitigate. Part of that risk was obviously that Russia could create a geopolitical crisis that would result in American companies losing access to those fuel services. Every firm knew the risks and responded accordingly, and those that moderated those risks by finding alternative suppliers or minimizing their exposure to Russian fuel should be rewarded. Those that did not should suffer the consequences. Indeed, many American utilities decided that the lower costs were worth that risk.

But the full story is not quite as simple as some companies making poor risk assessments about the downsides of using Russian fuel services. Unfortunately, American policymakers have been distorting commercial nuclear fuel markets for decades, and that has in large part created the conditions for reliance on Russian fuel. You could even argue that government policy created America's dependence on Russian uranium on purpose.

To understand this claim, you must first understand that America's nuclear industry rose out of the imperative to win the Cold War. Nuclear technology was born out of two national security efforts: the Manhattan Project and the desire to build nuclear-propelled submarines. In its quest to ensure that the United States would stay at the forefront of nuclear technology, the government helped create a commercial nuclear sector, of which power production was a key component. This led to a series of public–private partnerships with the goal of moving peaceful nuclear technology away from government and into the private economy. Though government remained a partner in the early years, nuclear energy's initial expansion was primarily driven by market forces. This approach worked well at first, with approximately one hundred new reactors being built across the country during the initial wave of development in the 1960s and 1970s.

Unfortunately, Washington never fully removed itself from the nuclear business, and as we've seen elsewhere in the book, this has resulted in an industry that has never reached its full potential. Government hasn't given it a predictable and reasonably regulated playing field where it could succeed or fail on its own merits. It never actually stepped aside to allow an independent, vibrant, and competitive commercial nuclear industry to emerge.

We can look at the failed effort to privatize enrichment services to see exactly how this played out. The US government started enriching uranium for national security purposes in the 1940s. In the 1960s, it began selling enriched uranium to commercial reactors around the world, and in 1992, it took the first steps toward privatizing this domestic enrichment infrastructure by passing the Energy Policy Act, which created the United States Enrichment Corporation, or USEC. Six years later, in 1998, the government sold USEC through an initial public offering.

That all sounds good, right? The government helped to build a critical industry at a key moment in world history, and the company it created found a global market for its product and was also able to advance the idea of peaceful nuclear power worldwide. And at the right moment, the government stepped back and privatized the company through an IPO. A success story by any measure.

The problem is that this was not the end of Washington's involvement with domestic enrichment. Instead of allowing a domestic industry to emerge from market supply and demand, the government continued to intervene in the enrichment business through a series of direct and indirect taxpayer-supported projects. The most important of these was a 1993 US-Russia agreement called the Megatons to Megawatts Program, which aimed to reduce Russia's huge stockpile of weapons-grade enriched uranium by converting it to fuel-grade uranium and selling it to American commercial nuclear reactors. By all accounts, the program was an arms-control success, converting 500 metric tons of weapons-grade uranium into 14,000 metric tons of commercial-grade uranium for nuclear fuel—approximately half the fuel needed for all American reactors for nearly two decades.[29]

Again, this sounds like a good news story. But the Megatons to Megawatts Program also eliminated the need for domestic enrichment capacity, and in 2013, America's last and inefficient gaseous diffusion enrichment plant in Paducah, Kentucky, was closed. Since then, the US has become more dependent on Rosatom, even as Putin has become more belligerent and aggressive.

The Megatons to Megawatts Program was supposed to run for twenty years. Many anticipated that in that time, there would be an expansion of US commercial nuclear power, and indeed by 2012, electric companies had announced plans to build nearly thirty new reactors that would all need enriched uranium fuel. But several market developments, including low growth in the demand for electricity, low natural gas prices, and generous government subsidies and mandates for use of wind and solar led to the cancellation of all but one of the new reactor projects. The same factors caused plant owners to announce plans to shut down a dozen operating reactors. The one exception was a uranium enrichment centrifuge facility in New Mexico, built by the European consortium Urenco. Today, that is the sole enrichment facility in the US.

WASHINGTON FINALLY RESPONDS

Shortly after the start of the Ukraine war, Senator John Barrasso (R-WY), ranking member of the Senate Committee on Energy and Natural Resources, introduced legislation (S. 3856) to ban imports of Russian uranium.[30] Unfortunately, legislative efforts like this move slowly. It took nearly two years for Congress and the administration to make meaningful progress.

In May 2024, President Biden signed the Prohibiting Russian Uranium Imports Act into law, which was a massive step in the right direction, even though it wasn't a complete or immediate ban on Russian imports.[31] But it did ban the import of low-enriched uranium from Russia or any Russian entity. This is an obvious foundation on which any ban must be built.

It also banned LEU that "is determined to have been exchanged with, swapped for, or otherwise obtained" in an effort to circumvent the ban. This is a critical piece of any meaningful prohibition because of the global nature of uranium markets, where uranium fuel often changes hands between producers, users, brokers, and dealers, providing opportunities to undermine the ban.

Finally, the ban lasts until 2040. This is critical. Expanding uranium enrichment infrastructure is time-consuming and expensive, and no private investor would expand capacity to make up for Russian supply based on a ban that is not long term and guaranteed. It should be noted, though, that growing demand for non-Russian uranium fuel had already prompted America's sole domestically located commercial enricher to begin expanding capacity even before the ban.[32] Nonetheless, a long-term ban on Russian imports provides the market certainty required to justify a broader expansion in the sector.

There are, however, potential stumbling blocks that may prevent the ban from maximizing its impact. One is abuse of the waiver process that the act establishes to allow some Russian imports through 2027 under certain conditions. While a waiver process may be necessary in specific cases, waivers must not be granted unless there is no alternative source for the LEU. One such waiver has already been granted. Unfortunately, at the time of this writing, there is little information available about how the waiver request meets the law's requirements. To ensure the act's integrity, Congress should demand a full justification from the Department of Energy regarding its decision to grant any waivers.

At the same time, Congress and the president must resist the urge to use subsidies and other taxpayer-supported schemes to drive the industry forward. Washington's intervention in uranium markets is at least partially to blame for the nation's current predicament. Instead, the focus should be on streamlining the process to expand, build, and start operating new enrichment plants. Given the industry's safety record, the private sector should be allowed to expand that effort with as little interference from Washington as possible. The executive branch also needs to embrace mining. Opening new mines in the United States is extremely difficult, and policies like Biden's decision to take domestic supplies out of service prevent uranium markets from responding to foreign supply disruptions.[33]

But if properly implemented, a ban on Russian uranium imports can be imposed with minimal impacts on supply while also avoiding major adverse economic impacts on utilities, their customers, and the broader American economy. To achieve this, however, policymakers must resist both their own temptation to intervene in markets through subsidies and the calls from some in the industry to do the same.

At the moment, unfortunately, they are failing miserably. Certain elements of the nuclear sector opposed banning Russian uranium in the months following the country's invasion of Ukraine because, they argued, that Russian uranium was necessary to maintain US electric system reliability and low prices. As a continued reliance on Russian fuel became politically untenable, arguments shifted toward securing public funding to subsidize domestic fuel-cycle expansion. Washington was happy to comply.

Instead of providing a predictable, stable, and affordable regulatory environment and allowing markets to grow naturally, Washington, and the Department of Energy in particular, is once again intervening with billions of taxpayer dollars to manage the growth of a commercial sector.[34] Like nearly all government spending meant to build commercial enterprises, this effort is unnecessary and will likely fail. It will either subsidize investment that would have occurred naturally or encourage malinvestment that is not supported by market forces. And it's unfair to the firms that took the initiative to expand capacity without Washington's help.

The fact is that uranium fuel markets fall short, especially in the US, not because they lack subsidies but because bad policy misaligns incentives, creates barriers to success, and increases financial risk, all of which undermines private investment. Rather than mitigate that risk through subsidies, government should focus on eliminating the risks that it creates and allow market forces to drive the industry forward.

The uranium fuel industry is not starting from square one. Uranium ore is globally available, and there are significant non-Russian providers of conversion services. In the US, a uranium conversion plant in Metropolis, Illinois, was the primary supplier to US reactors for decades. As mentioned earlier, it was shut down in 2017 because of excess global capacity and low demand, but the NRC has granted its request to renew its operating license to March 2060.[35] In addition to this domestic source, there are conversion facilities in France, Canada, and China.

While the United States certainly does not want to exchange a dependence on Russia for a dependence on China, the fact is that growing Chinese contributions to the commercial nuclear industry will play an increasingly significant role in meeting global demand. This will free up resources for the United States and its allies.

Replacing Russian uranium enrichment is still a challenge, but it can be overcome. In 2023, US companies purchased 15 million SWU from eight sellers under uranium enrichment contracts.[36] Twenty-eight percent of this SWU was provided domestically. Russia provided 27 percent, and friendly nations like the United Kingdom, Germany, and the Netherlands provided most of the rest.

Based on 2023 numbers, the US would need to make up about a quarter of its enriched uranium needs to cut Russia out of the picture. That's a large but not impossible amount, especially because nuclear operators have 141.7 million pounds of uranium concentrate equivalent on hand. Of course, not all of this inventory is enriched uranium, but some portion of it is, and that provides an additional cushion. The US has enough enriched uranium to meet its near-term needs, but without new Western enrichment capacity, global enrichment needs will exceed annual non-Russian supply by up to 20 percent through 2038.[37]

This underscores the point that waivers under the ban should be issued very conservatively.

AND THEN THERE IS THORIUM

There is another way to minimize our dependence on foreign uranium, and that is to diversify the fuel we use for commercial reactors. Though the current nuclear industry is largely fueled by uranium, other reactor technologies use different fuel cycles, such as thorium, and some can even derive power from used nuclear fuel. Let's look closer at thorium, which is three times more plentiful in the earth's crust than uranium and has some other interesting properties as well.[38]

A thorium reactor produces uranium-233 (U-233), which is a form that doesn't exist naturally. It exists only when it is produced through nuclear fission using thorium. When a thorium atom absorbs a neutron, it becomes uranium-233, which is valuable for two major reasons: (1) it can be used to generate energy from natural thorium, essentially forever because it can breed more fuel than it burns inside the reactor,[39] and (2) a small amount of it decays into special medical radioisotopes that are incredibly effective in fighting cancer.[40] And there is more. Thorium reactors produce less nuclear waste than uranium reactors, and some can even use existing nuclear waste as fuel.

Here's where it gets even more interesting. The United States is the only country that has much U-233. American scientists realized early on in their nuclear weapons research that U-233 was not practical for making bombs, so they shelved it, but not before producing substantial quantities of it. Nuclear engineers, however, understood that the material would be great for fueling power reactors. In fact, an experimental reactor at the Shippingport plant in Pennsylvania not only operated successfully using U-233 in the 1970s, but was shown to breed fuel as part of its operations.[41] Thorium has been used to partially fuel two other commercial reactors over the years as well, though it was a different fuel cycle than the one at Shippingport. These included high-temperature gas-cooled reactors at Fort Saint Vrain in Colorado and Peach Bottom in Pennsylvania, both of which operated commercially at points in the 1960s, 1970s, and 1980s.[42]

This was a major technological milestone that could provide the basis for a nearly limitless energy resource that America's lawmakers seem to have forgotten about. Not only has little been done on the policy front to invite thorium-based reactors into the market, but plans are in place to render America's stockpile of U-233 essentially useless so it can be thrown away.

That's a terrible idea for a lot of reasons, but two are abundantly clear. The first is that U-233 should be used for what makes obvious sense, which is to generate energy from thorium forever, and a new generation of nuclear reactors based on molten salts that use thorium make this possible. American companies are working on this technology today, in competition with state-owned Chinese efforts. Even though China plans to spend $440 billion over the next fifteen years on its nuclear development strategy, which includes thorium research,[43] America has a distinct advantage in its stockpile of uranium-233, which China doesn't have. The second reason is that U-233 makes medicines that fight cancer. By destroying our stockpile of U-233, we are essentially making the development of these treatments more difficult. This make no sense and it's also expensive. In 2018, the US spent $33 million to $52 million on the program to ultimately dispose of its U-233 stockpile.[44]

Fixing America's nuclear fuel markets is not just about banning Russian imports and producing more uranium fuel domestically but also about diversifying the fuels we use for commercial reactors. Thorium reactors are a proven technology, and private firms are working hard to develop commercial applications. If the United States is serious about nuclear energy and energy independence, then the federal government needs to recognize the important contribution that thorium can provide and remove policy and regulatory biases against the technology.

BEYOND RUSSIAN URANIUM

A properly implemented ban on Russian imports will encourage the development and expansion of US nuclear infrastructure and strengthen collaboration and trade relationships with allies and reliable partners. While the Prohibiting Russian Uranium Imports Act is very good first step, more should be done, including the following:

- **The executive branch should encourage allies to implement their own restrictions.** To maximize the foreign policy benefits, as well as provide additional long-term certainty about future market opportunity, the US should lead an international effort against Russian energy exports. This would further insulate America and her allies from Russia's ability to use energy as a weapon and allow US firms to expand their capability to win foreign business and serve foreign customers. This collaboration will also identify areas of fuel supply chains where more investment is needed.

- **Congress should direct the administration to conduct an exhaustive audit of all potential uranium resources that are not earmarked for national security purposes.** While much of the government's excess uranium has been dispositioned over the past decade or so, some remains. Given the importance of ending US reliance on Russian uranium and the fuel supply disruptions that could result from a ban, these remaining available uranium resources should be identified and set aside to fill any supply gaps.
- **Congress and the administration should recognize the thorium fuel cycle's potential in meeting its energy goals.** Thorium fuel represents an important opportunity to diversify America's nuclear fuel mix, to establish global leadership of an emerging technology, and to develop lifesaving cancer treatments. But taking advantage of these opportunities will require that certain actions be taken. First, Congress should immediately stop all uranium-233 disposal activities. The stockpile of U-233 is a strategic asset that should be protected for future use. Second, efforts at the NRC to modernize the regulatory process to better accommodate advanced reactors should take specific care to ensure that newly established processes do not bias against thorium-based reactors. And finally, reforms to nuclear waste policy, which will be discussed in the next chapter, should recognize the unique role that thorium reactors can play in managing America's spent nuclear fuel.
- **Government agencies should streamline the process for US companies to export peaceful nuclear technology.** A critical component of convincing allies to put their own limits on Russian fuel services is ensuring that American products and services are available to fill their commercial nuclear needs.

Unfortunately, our commercial nuclear companies are at a severe disadvantage compared to firms in places like Russia and China because of our overly bureaucratic system for exports. To better compete, America must streamline its system, and the Departments of Commerce, Energy, and State all have roles to play in that, as does the Nuclear Regulatory Commission. To that end, Congress should review the current agreements (including the so-called 123 Agreements[45]), laws, regulations, and practices that govern commercial nuclear exports, and then modernize, reduce, and simplify them. This will result in a predictable and efficient process that allows US nuclear firms to compete with state-owned entities from Russia and China.

- **Congress and the administration should remove barriers to building nuclear infrastructure facilities in the United States.** Perhaps the most important thing that the federal government could do to promote an expansion of uranium enrichment and end dependence on foreign suppliers is to fix the policy barriers that make it difficult to build new nuclear facilities. To make it easier and more cost-effective to replace the banned Russian uranium, Congress and the administration should significantly lighten the regulatory burden on building new US enrichment capacity and nuclear power plants, especially those using NRC-certified designs. Better regulation will enable the nuclear industry to make the significant and long-term investments required to expand enrichment capacity, strengthen cooperation with allies and other reliable trading partners, and develop new supply chains outside of Russia.

Some in the US nuclear industry feared that a uranium ban would cause higher fuel costs or shortages. This was a legitimate concern because our nuclear fuel infrastructure has suffered from neglect, short-sighted policies, and underinvestment. But a thoughtfully implemented ban, coupled with proper policy reforms, will mitigate the risks and provide the right incentives to move America past its reliance on Russian uranium.

Banning Russian uranium imports is a good beginning to righting America's nuclear policy. But it must be done prudently. We can't ignore the consequences of past policy decisions on the current industry or make the same mistakes again. The president and Congress should commit to creating an efficient and stable political and regulatory regime that will allow buyers and sellers to shape an industry that serves consumers and the national interest and encourages investment in a nuclear future free of a reliance on Russia.

CHAPTER 10

THE ATOMIC OPPORTUNITY OF NUCLEAR WASTE

To create power, the fuel used in most of today's reactors must contain 3 to 5 percent fissionable uranium. Once the fissionable uranium reaches around 1 percent, the fuel must be replaced. But this spent fuel generally retains about 95 percent of the uranium it started with (albeit at a lower enrichment level), and with the right technology, that uranium can be recycled for future use as nuclear fuel or other commercial purposes—more on that later. The sad thing is that the United States developed the technology to recapture that energy decades ago, but thanks largely to fearmongering from the antinuclear movement in the 1970s, the government barred its commercial use in 1977. The US has had a virtual moratorium on commercial nuclear fuel recycling ever since.

Other countries have not taken such a backward approach. France, for example, whose fifty-six reactors generate 80 percent of its electricity, has been safely recycling nuclear fuel to access uranium and plutonium for fuel since the start of its nuclear power program in the 1970s. When used fuel is removed from French reactors, it's packed in containers and safely shipped via train and road to a reprocessing facility in La Hague, on the northwestern coast. There, the energy-producing uranium and plutonium are separated from the other waste and made into new fuel that can be used again. The entire process adds about 5.5 percent in costs for the French.[1] Over the last decade, recycling costs have been reduced by 40 percent.[2] France has reprocessed over 23,000 tonnes of nuclear waste since the mid-1960s, which is about fourteen years' worth of nuclear fuel for the country.[3]

France's success has sparked plenty of interest abroad, and a French company has already helped Japan with its reprocessing facility and is currently examining the feasibility of building a similar plant in China. And if interest in nuclear power continues to expand in the US, France will certainly be part of that conversation, as it has been during past instances of heightened nuclear interest. The British, Japanese, Indians, and Russians have all engaged in some level of reprocessing, and South Korea is looking to do the same. Of course, some highly radioactive waste is still produced. But the French deal with it effectively—placing some in short-term storage and preparing the rest for long-term storage in a future geologic repository, much like the planned US repository at Yucca Mountain (more on that later).

In La Hague's many decades of operation, it has recycled fuel without incident: no successful terrorist attacks, no evil masterminds stealing uranium, no accidental explosions. But all is not perfect. Most critically, the French still do not have a permanent geologic storage facility. What they do have, though, is the ability to handle used nuclear fuel and advance their program without being held hostage to the politics of geologic storage.

While the French model, which is essentially state ownership and relies on a single technology, may not be appropriate for the United States, it does demonstrate that used-fuel recycling can be integrated into a comprehensive nuclear energy strategy. The US should learn from the French experience while also seeking to develop a uniquely American model that is innovative, agile, technologically diverse, and led by the private sector.

For the United States to reestablish itself as global leader in commercial nuclear power, it must modernize its approach to nuclear waste management and disposal. Without such a commitment, it will remain at a competitive disadvantage to those nations that can offer complete fuel cycle services.

NUCLEAR WASTE POLICY IN THE UNITED STATES

The Nuclear Waste Policy Act of 1982, as amended, attempted to establish a comprehensive disposal strategy for high-level nuclear waste.[4] It charged the federal government with disposing of used nuclear fuel and created a structure through which users of nuclear energy would pay a set fee for that service. These payments would go to something called the Nuclear Waste Fund, which the federal government could access through congressional appropriations to pay for disposal activities. The waste was supposed to be permanently stored at a repository at Yucca Mountain, Nevada.

By any measure, this strategy has failed. To date, the government has spent around $11 billion without opening the repository or receiving any commercial waste, other than what was in the reactor after the accident at Three Mile Island. Because it has refused to take possession of the used fuel—despite the act's January 31, 1998, deadline—the government is liable to nuclear plant operators for the costs associated with holding spent fuel. As of September 2023, taxpayers had paid out around $10.6 billion to utilities that sued the government for breach of contract.[5] The Department of Energy claims that total outstanding liability exceeds $30 billion. Nuclear industry experts say it's significantly more than that and will come in at around $50 billion.[6]

An interesting and important side note is that this money comes from the federal Judgment Fund, which is a permanent and indefinite fund to settle financial judgments against the federal government. In essence, the Department of Energy entered into contracts with private parties to dispose of waste, collected money from electricity ratepayers to fund those disposal activities, failed to actually do what it was contractually obligated to do, and then wriggled free of the consequences and left taxpayers holding the bag.

Meanwhile, the 1982 act has disincentivized the firms that produce waste and would benefit most from its efficient disposal from finding more workable alternatives. And the government's inability to fulfill its legal obligations has often been cited as a significant obstacle to building new nuclear power plants. Now is the time to break the impasse over managing the nation's used nuclear fuel.

Currently, the United States has around 90,000 tons of high-level nuclear waste stored across eighty sites, and its commercial reactors produce approximately 2,000 tons of used fuel every year.[7] Through the Nuclear Waste Policy Act, the Yucca Mountain repository's capacity is statutorily limited to 70,000 tons of waste, with 63,000 tons allocated to commercial waste and 7,000 tons to the Department of Energy. In other words, if Yucca becomes operational tomorrow, it will have already reached its limit unless the defined capacity is changed.

These are artificial limitations that Congress set without regard to Yucca's actual capacity, which is much larger than the prescribed limit. According to the Department of Energy, Yucca Mountain has enough room to hold all of the spent fuel produced by America's current fleet of reactors.[8] Other studies have found that Yucca could safely hold much more waste than that.[9]

So why did Congress set such a low limit? At the time, there was a desire to build a second repository, and one way to make that happen was to set an artificially low ceiling on the initial repository. The statutory limits should have been lifted once lawmakers decided that Yucca would be the only repository built, but even with an expanded capacity of, say, 120,000 tons, the site could hold only a few more years' worth of America's nuclear waste, assuming the US significantly increases its nuclear power production. According to one analysis, the country's current reactors would generate enough used fuel to fill a 70,000-ton Yucca right away. If nuclear power increases by an average of 1.8 percent annually starting this year, a 120,000-ton Yucca will have reached its limit by 2037. At that growth rate, the US would need more than three Yucca Mountains by the turn of the century.

Given what's happened with Yucca, it seems foolish to rely on the promise of future repositories. Foolish and unnecessary. With the right mix of alternative practices, such as storage and recycling, Yucca or another deep geological repository could last virtually indefinitely. There are already many other available technologies that could help solve America's nuclear waste problem once and for all.

OVERHAULING USED-FUEL MANAGEMENT

A sustained rebirth of nuclear energy in the US isn't possible without a rational and functional program for the disposal and management of nuclear waste and spent fuel. New nuclear plants could last as long as one hundred years, but to reap the benefits of such a long-term investment, they must have a sustainable avenue to deal with spent nuclear fuel. Establishing a practical pathway for waste disposal is essential to ensuring long-term plant operations, and such a pathway would also mitigate much of the political risk associated with nuclear power.

The antinuclear movement has long recognized this fact, and this is why they've spent decades demonizing anything to do with nuclear waste and any activity associated with it. Indeed, they have worked specifically to amplify and weaponize this political risk by propagating the narrative that nuclear waste management and disposal is unsafe. This would be problematic regardless of who is responsible for it, but as long as the federal government is in charge of disposing of waste, the process will be subject to the political arena, which has made it extremely vulnerable to antinuclear propaganda. And because the federal government is the only entity in the entire nuclear food chain with any incentive to fix the problem, this strategy has been extremely effective. Whether it's because of political pressure from the antinuclear movement, ignorance, or hubris, the federal government has done little to move the nuclear waste ball forward in any meaningful way.

That said, the Department of Energy has opened the door to reform at different points over the years, but it has never seriously considered a divestment strategy, where the responsibility for managing and disposing of spent nuclear fuel is returned to the private sector. Administrations come and go, but the inflexible rules and bureaucracies that govern waste management seem to endure forever, making it impossible for Washington to respond effectively to a rapidly changing industry. When it tries to, it often acts in ways that make no business sense, are inconsistent with the actual state of the industry, and are more concerned with political optics than anything else.

Many of these efforts result in large government programs. While some of these programs have had near-term benefits—they demonstrate political support for nuclear power, encourage private and public research and development, and advance the nuclear industry—they inevitably do more harm than good. They are run inefficiently, are sometimes not completed, cost the taxpayers billions of dollars, and are often not economically rational. Furthermore, they frequently lack long-term planning, which leads to unsustainable programs that provide fodder for antinuclear critics and ultimately set industry back by discouraging progress in the private sector.

When it comes specifically to nuclear waste management, there are three fundamental problems:

1. **No long-term geologic storage.** Deep geologic storage like that proposed for Yucca Mountain is a safe long-term solution to dealing with spent fuel and thus is critical to any comprehensive nuclear waste management plan. But despite having collected and spent billions of dollars in pursuit of this plan, the US still has no functional geologic repository for commercial waste anywhere in the country.
2. **No incentives for the private sector to get creative when it comes to waste management.** Private plant operators produce nuclear waste, but under the current law, the federal government is responsible for managing and disposing of it. This means the firms that produce waste (or spent fuel) have no interest in how the waste is managed and no reason to come up with alternative approaches. At the same time, Washington has proved unable to implement anything close to a workable solution. This was a predictable outcome given a structure that fundamentally misaligns incentives, responsibilities, and authorities. The nuclear industry, which is fully capable of running safe nuclear power plants, is qualified to manage its own waste and should have the responsibility to do so.
3. **No specific price for specific services rendered.** Under the current system, nuclear utilities produce waste and then pay

the federal government a flat fee for an undefined (and undelivered) service. Accurate pricing is critical to any efficient marketplace. Prices provide both suppliers and purchasers with a critical data point by which to determine the attractiveness of a product or service, and they give potential competitors the information needed to introduce alternatives.

To at least some extent, most observers recognize that America's system for nuclear waste management is flawed and should not reside with the Department of Energy. There is also broad agreement that whatever plan is ultimately developed, it cannot be devoid of market forces.[10] In 2012, to try to move America's nuclear waste management strategy forward, President Obama asked the energy department to put together the Blue Ribbon Commission on America's Nuclear Future (BRC).[11] The BRC introduced several important reforms and remains the gold standard in this area, but its recommendations generally accept the idea that a centralized entity, established and empowered by the federal government, should be responsible for nuclear waste management.

The centerpiece of the BRC's recommendations is a proposal to establish a federal corporation "dedicated solely to implementing the waste management program and empowered with the authority and resources to succeed." While the general proposition could help transition the US toward a more market-based system, the organization envisioned by the BRC will not result in a dynamic, innovative, and competitive nuclear waste management industry because it maintains the existing system's basic underpinnings: a government-based entity would still be responsible for waste management and disposal, relieving waste producers of all responsibility, and there would be little direct connection between pricing and services rendered. To result in real reform, the new organization cannot be a creation of government and generally maintain the same responsibilities currently held by the Department of Energy. Simply moving a function from one government agency to another (even if the new agency is called a federal corporation) without changing the system fundamentals only creates the perception of action while perpetuating existing deficiencies.

This approach assumes that the basic premise of the current system—that nuclear waste management and disposal should be within the purview of the federal government—is correct. It essentially blames systemic problems on a misplaced federal bureaucracy when the actual problem is relegating a commercial activity to a government agency. The goal should be to transfer the responsibility for nuclear waste management and disposal away from Washington and toward the private sector.

But even though a federal agency is not the appropriate entity to permanently take on nuclear fuel management, it could still be a key component in facilitating a transition to private-sector responsibility. This process is not going to happen overnight because the federal government is contractually obligated to deal with the disposal of nuclear waste produced at existing plants, and the nuclear industry, through fees levied on power users, has already paid a lot of money for that service. A federal corporation, limited in scope, could be the correct entity to take responsibility for disposing of that waste.

Such a federal corporation should have two basic responsibilities. First, it should choose a site for a geologic repository. If the repository is to be located at Yucca Mountain, as the current law stipulates, then the federal corporation should take over from the Department of Energy to complete the construction and operation permit application. Once issued, the permit should be transferred to a non-federal entity to construct and operate the facility. But if Yucca is deemed technically deficient, the corporation should oversee the selection of a new location. The permit application should be prepared by whichever entity will eventually construct and operate the facility.

The corporation's second responsibility should be to assure proper disposal of the nuclear waste for which the federal government is currently responsible. The corporation should be given access to the Nuclear Waste Fund, not to finance the construction or operation of any waste management facilities but to pay for waste management services from the private sector. These services could include placing waste in a future repository, using advanced reactors or a technical process like reprocessing to manage the waste, or even transferring the waste to a third party that sees future value in it. This would let the federal government meet its existing contractual and regulatory responsibilities while also allowing an eventual transfer to the private sector. Most important, it would create a significant market demand for privately offered waste management services like storage, transportation, and processing. Businesses would naturally emerge to meet this demand, which would then be available for future private waste management operations.

Finally, the transitional federal corporation must be mission-specific, and its creation must be accompanied by a dissolution plan. Once its two responsibilities are met, it should be either privatized or abolished.

TRANSITIONING TO A NEW SYSTEM

All geologic repositories should be operated by non-federal entities. The management organizations could be private, for-profit, non-profit, state-based, or a combination of any of these. One of their basic responsibilities would be to set market prices for waste emplacement, taking into consideration waste characteristics such as heat load, toxicity, and volume, as well as available repository space. Waste producers would then have different variables to consider when deciding which fuels to purchase and which nuclear technologies to use because these decisions would affect how they ultimately manage their waste. For some nuclear plant operators, it could be most cost-effective to place waste directly in a repository, while others might find interim storage or another process to be more economical. Market-based price signals would encourage new services and technologies, such as reprocessing and small nuclear reactors that have different waste streams or could even be used to dispose of existing waste.

The opportunity presented by transferring responsibility for spent fuel to the private sector does not end with finding better ways to store it or extracting materials for future fuel. Spent fuel also contains a myriad of other valuable resources, including rare earth metals and medically useful isotopes.[12] We're already seeing new businesses built around the idea that spent nuclear fuel is a prized asset that could be worth billions. But the value will never be realized until it is exposed to a free market, and to achieve that, we must shift the responsibility for managing and disposing of future fuel to those who produce it.

Moving forward, all nuclear utilities should be made responsible for the waste they produce. This responsibility should be accompanied by a repeal of the Nuclear Waste Fee paid to the federal government for waste disposal. Utilities would then bear the responsibility and also have the freedom to choose how best to manage their waste. The government's role would be to ensure that private waste management activities meet adequate regulatory standards. In essence, waste management would be treated like all other aspects of the nuclear industry. The federal government isn't responsible for getting the fuel to the reactor, and it shouldn't be responsible for taking it away.

Transitioning to such a system will likely cause anxiety among those who work in the nuclear industry—they should rightfully be skeptical that state and federal bureaucrats will stay out of the way of building the necessary infrastructure to manage spent fuel. To address this and also ensure that nuclear waste producers have access to waste management services, the federal corporation could be permitted, for a fee, to take title of waste produced under the new system until a private marketplace emerges. The federal corporation could also be permitted to broker waste management services for private industry—again, only until a viable waste management market becomes established. It may be the case that as the corporation gains experience and establishes relationships with waste management providers, it can negotiate better terms based on volume or other variables. Or perhaps waste producers will find the convenience of contracting with the federal corporation to be worth a premium.

Waste producers would not be obligated to seek waste management services through the federal corporation, though. This service would be available only as long as the federal corporation is carrying out its chartered mission, and it would not justify its existence as a public entity beyond those specified responsibilities. However, one can imagine a business case where brokering such services could provide the basis for future privatization. Ultimately, while such an arrangement is not necessary, it does provide an additional transition step toward the new market-based system.

Once the federal corporation is dissolved, the government should have two roles. First, it should set broad regulatory guidelines for waste management, just as it does for other parts of the nuclear industry. Second, it should take final legal title of any future geologic repositories that may be built, as well as their contents once they are decommissioned (at which point the repositories would be permanently sealed). Any additional costs associated with the repository, which will be minimal, should remain the responsibility of those whose waste was placed there.

FINANCING NUCLEAR WASTE MANAGEMENT AND DISPOSAL

When it comes to financing, there is little debate around the problems with the current system—it doesn't work as intended, and continuing to collect fees for services not rendered is patently unfair. It must be noted, though, that courts have ruled that utilities aren't responsible for paying the Nuclear Waste Fee until the Department of Energy establishes a program to manage waste. Nonetheless, the point still stands because the old system will eventually be reinstated unless reforms are enacted.

There is also a broad recognition that government accounting rules (whereby Congress must appropriate monies that were paid into the Nuclear Waste Fund) make gaining access to financial resources extraordinarily difficult. Subjecting any business proposition to congressional politics is a recipe for failure. That this business is nuclear energy makes things exponentially more difficult. As long as nuclear waste policy relies on the inherently inefficient and unpredictable congressional appropriations process, it will never succeed.

Separating finance issues from larger organizational issues is impossible. The two are inherently related. How nuclear waste activities are financed will ultimately depend on who is doing them. Therefore, any rational financing scheme must be developed in hand with larger organizational reform. If we accept the general proposition that the federal government should remain responsible for nuclear waste management, any finance reform will continue to be subject to the political process. So no matter what protections are put in place to ensure access to the money, the same inefficiencies that routinely result from federal control would ultimately resurface.

The BRC did provide a framework for building a more market-based, economically rational system. Indeed, it introduced some elements that are critical to a sustainable waste management system. But instead of trying to modify the current system to work better (which most nuclear waste financing reforms do), we should transition to a new model for financing waste management while ensuring that existing resources are used for their intended purposes. To achieve this transition, Congress should do three things:

1. **Immediately begin transferring the Nuclear Waste Fund to the new organization.** Whoever is ultimately responsible for waste management and disposal must be given access to the Nuclear Waste Fund. The entirety of the fund should be appropriated to the new organization over a specified period—say, ten years. The new organization would then be responsible for financing the disposal of all waste attached to the Nuclear Waste Fund proceeds. It should be left to the new organization to decide how to use the money and dispose of the waste. The only requirement should be to dispose of the waste attached to those dollars or to pay some other firm or entity to take on that responsibility.
2. **Mandate the creation of utility-specific or plant-specific escrow accounts to fund waste management activities.** An innovative concept outlined in the BRC report is to create escrow accounts, held by an independent third party, into which nuclear waste fees are paid. While that model is inconsistent with a true market-based approach, the BRC deserves credit for normalizing the escrow account idea. A better model would mandate that nuclear utilities place in escrow enough money to dispose of whatever waste they are storing on-site. No funds would ever go to the US Treasury or be under government control, or be required to be held by any

third party, and congressional appropriators would have no role. Utilities would simply pay for waste management and disposal services on an as-needed basis out of funds accrued in the escrow accounts. This approach would benefit utilities by ensuring they have access to the funds, and it would protect the American taxpayer by ensuring adequate funds are available even if a plant goes out of business.

3. **Repeal the fee paid to the federal government for future waste disposal services.** Under these reforms, existing nuclear waste disposal would be financed through existing waste fund money and future disposal through the privately held escrow accounts funded as reactor operators produce nuclear waste/spent fuel. That means there would be no need for operators to continue paying the Nuclear Waste Fee to the federal government. Even though the fee has been suspended since 2014, when courts ordered the Department of Energy to cease collecting it due to the department's failure to carry out its responsibilities, Congress should prioritize repealing the fee once and for all.

A NEW BEGINNING FOR YUCCA MOUNTAIN

The political establishment can continue to oppose a permanent repository at Yucca Mountain as prescribed in the Nuclear Waste Policy Act and simultaneously offer a better solution that would benefit all parties and perhaps result in opening the repository under different conditions. Within the context of the reforms offered above, it's possible to imagine a transaction that would include accepting title to the spent nuclear fuel that was produced in association with payments into the Nuclear Waste Fund in exchange for the entirety of the fund—the balance of which is currently around $50 billion. In other words, support for a repository at Yucca Mountain could grow substantially in Nevada if the idea were that the entirety of the Nuclear Waste Fund would be the payment for accepting the spent fuel.

Past surveys have shown that Nevadans are at least open to Yucca. Officials in Nye County, where the repository would be located, have long supported the project because of the potential economic benefits, including the possibility of thousands of new high-paying jobs.[13] A program that places Nevada in control of the future of Yucca Mountain could be successful. Under such a program, the state could negotiate directly with nuclear waste producers to come up with a mutually beneficial arrangement that sets a price for emplacing future spent fuel.

Unfortunately, a lack of innovative thinking from policymakers has essentially made Yucca Mountain a binary choice: support or oppose. A plan that could move the repository forward would require the NRC to finish its review of the DOE's permit application. The NRC's September 2008 docketing of the DOE's application to construct the repository set in motion a three-year, two-track review process that was supposed to culminate in a yes-or-no decision. Unfortunately, the Obama administration effectively terminated the Yucca Mountain program, an act that former NRC commissioner Kristin Svinicki described at the time as "grossly premature."[14] The permit process should be restarted and completed.

If the NRC issues the permit, policy changes (including legislative ones, as necessary) should define a process whereby the license to operate Yucca is held by a non-DOE third party, such as a private-sector nonprofit or even the state of Nevada itself. The new permit holder could then negotiate a workable path forward that would fully represent the interests of all parties. This process was absent from the original decision to name Yucca the waste repository site.

For the purposes of this discussion, let's assume that an entity closely associated with the state of Nevada is the one holding the permit. It could be that this entity decides it still doesn't want the repository or the financial windfall that would come with it. In that case, the entity could simply hold on to the permit and never act on it. But this would then create an opportunity for some other community to take the deal. This scenario could ultimately lead to Yucca being opened as a second repository. As the initial repository becomes operational, people will see that the alleged risks of a spent fuel repository are minimal and well worth the economic opportunity created. This could inspire Nevada to reconsider its resistance to a repository at some future point.

The long history of opposition to Yucca makes it difficult for many politicians to support the project, despite evidence of its technical soundness and general safety. This is unfortunate because Nevada could have benefited greatly from the economic impact of the facility. But this opportunity lost is also an opportunity gained. The United States and Nevada now have a chance to chart a new path forward. Such reform would help not only to establish a new industry in Nevada but also to bring nuclear power back in the US.

Either way, the current system for managing nuclear waste has been an abject failure. Though there is widespread acknowledgment that the system doesn't work, too many reform efforts do little more than tinker around the margins while leaving the underlying fundamentals in place. Such measures may create the perception of progress, but the same old issues will ultimately emerge. Even if the federal plan did work as intended, the United States would still lose the vast benefits of a competitive market-based system. Introducing market forces into the process, empowering the private sector to manage nuclear waste, and engaging communities in new and innovative ways will solve this problem, but the federal government will need to step aside and allow the private sector to take charge.

CHAPTER 11
A POLICY REVOLUTION

The federal government can't build an innovative, competitive, and economically independent commercial nuclear industry. America now has decades of proof that confirms this fact. Private enterprise, on the other hand, has the interests, expertise, and background to develop a cost-effective industry that is economically independent of government and competitive in domestic and international markets. This doesn't mean that Washington should have no role, but policymakers must recognize that government intervention as traditionally applied has not produced a viable nuclear industry.

The term "nuclear renaissance" is often used to describe the reemergence of nuclear power in the US and around the world. But a renaissance should not be the objective—nuclear energy has too much promise for humanity to think so small. What we need is a revolution.

Over the years, a lot of good ideas—for instance, on how to liberalize the commercial nuclear trade, how to update regulatory structures, and how to make it easier to finance reactor projects—have been put forward. These reforms are steps in the right direction, but they will only move nuclear forward on the margins. They accept the basic framework that governs how the nuclear industry operates in the United States today, and they rely primarily on the foundational statutes and regulations that have been in place since the dawn of the commercial nuclear era. As we've seen, the current system has led to massive cost escalations, the proliferation of subsidies, a loss of US global leadership, the prioritizing of politics over economics, and precisely two new reactors.

The time for tinkering around the margins has passed.

We need a revolution.

Policymakers should create an entirely new framework for commercial nuclear energy that is wholly rooted in free enterprise, innovation, and competition. This approach should not completely replace the current framework (yet), but it's long past time to fully free American commercial nuclear power—a mature and safe industry—to step into the twenty-first century.

As they build this new approach, policymakers should focus on seven key areas:

1. End all energy subsidies
2. Modernize low-dose radiation exposure standards
3. Broaden state authority to regulate nuclear power
4. Focus the NRC on areas where it has a competitive advantage or established expertise
5. Develop regulatory alternatives around private liability insurance requirements
6. Apply market principles to spent nuclear fuel policy reform
7. Fix nuclear fuel markets

Let's take a look at each of these seven areas in more detail.

END ALL ENERGY SUBSIDIES

Energy markets are massive. In the US, consumers spend around $1 trillion a year on energy, and the global market is ten times that.[1] Energy is a product that every one of the eight billion people on earth wants and needs, and each year, billions of dollars are spent researching and developing new energy sources. If ever there was a market that could function without government intervention, it is this one.

And yet, governments cannot help but intervene. They create entire narratives around scarcity, environmental harm, justice, cost, and public health to justify their policies. And when mountains of evidence show that their interventions often lead to fewer options at higher costs with suboptimal environmental outcomes, they just keep adding more. Worse, each subsidy or intervention that's layered on exacerbates the market inefficiencies, which leads to even poorer outcomes. Of course, the politicians and bureaucrats then argue that fixing these problems requires more subsidies and interventions. Ultimately, this is all a manufactured narrative to justify whatever agenda is being pursued or special interest is being mollified.

Some will argue that subsidies for one energy source are justified because some other energy source currently receives subsidies or did so in the past. This is a non sequitur. One subsidy does not justify another. The truer and more persuasive argument should be that existing subsidies distort the market and result in bad outcomes for citizens, and thus should be removed. In the unlikely event that a subsidy or intervention is necessary, it should be judged on its own merit and nothing else.

Others will argue that subsidies are necessary to kick-start certain industries. Even if we ignore the fact that this argument is most often applied to energy sources that have been around for decades, if not centuries, it depends on the idea that somehow certain sources have a right to exist. They do not. Energy is a tool used to power society. How that energy is produced is almost irrelevant. It only matters that people and businesses have the energy they need at affordable prices without endangering public health and safety. That's it.

Perhaps you could argue that government interventions create value for citizens by resulting in more choice and better prices over time. But they do not. They perpetuate mediocrity and kill innovation. They create dependence and debt and rent-seeking. Some will point to this technology or that one and say those wouldn't exist without early subsidies. That is a vacuous argument as well, and it goes back to the seen and unseen argument that Bastiat taught. Sure, it might be true that a certain enterprise received a subsidy early on and became successful, but that success was because the enterprise produced something that people wanted. What we don't know is how that firm would have responded if the subsidy was never received. It likely would have figured out a way to succeed if its product was worth selling. What we also don't know is what other firms lost out because they couldn't compete with the subsidized firm, or what they could have done with the money the subsidized firm received.

The fact is that more than enough public treasure has been expended to kick-start, develop, or keep afloat enough energy firms and technologies. The first step in helping not just nuclear but all energy sources that have something important to contribute to society is to end all energy subsidies forever.

MODERNIZE LOW-DOSE RADIATION EXPOSURE STANDARDS

Regulating nuclear energy is primarily about protecting people and the environment against radiation exposure. The Environmental Protection Agency is responsible for establishing generally applicable, legally enforceable standards for that purpose. The agency is governed by a series of statutes that set standards for things like radiation emissions from everyday nuclear power operations and the safe disposal of spent nuclear fuel and uranium mill tailings.[2] These standards then inform federal regulatory agencies, like the Nuclear Regulatory Commission, as they fulfil their missions.

The EPA bases its risk assessments, regulatory limits, and guidance on the linear no-threshold (LNT) model. According to the EPA, LNT "assumes that the risk of cancer due to a low-dose exposure is proportional to a dose with no threshold. In other words, the LNT model assumes that ionizing radiation is always harmful and that there is no threshold amount below which radiation exposure is safe."[3] The NRC has affirmed its commitment to the LNT model as well.[4]

It's interesting to note that the EPA accepted the LNT model in 1975.[5] This is around the same time that nuclear energy costs began skyrocketing. That hardly seems a coincidence.

There is tremendous controversy around the legitimacy of LNT. Ample evidence demonstrates that low-dose radiation is not harmful to humans, and some research even shows that it can be beneficial.[6] This makes sense considering that we evolved in an environment where we are constantly exposed to low-dose radiation that emanates from natural sources in the earth and space.[7] Natural radiation varies from region to region, with some areas experiencing significantly higher levels of exposure. Research has shown that people who live in these regions do not experience higher cancer rates. One study concluded that "[n]either cancers nor early childhood deaths positively correlate with dose rates in regions with elevated natural background radiation."[8] Even the NRC recognizes that LNT may overestimate risks.[9]

But despite all this evidence, the EPA remains committed to the LNT model. Frustratingly, the agency recently undertook a process to determine whether it should continue using the model and based its conclusion not on independent science or a holistic review of evidence but on "recommendations by national and international authoritative bodies."[10] In other words, to justify its continued use of LNT, the EPA essentially relied on the conclusions of the same organizations that have been promoting LNT.

While politicians and bureaucrats like to use complex models and confusing academic theories to support their agendas, common sense and observable experience would tell us that we should not fear low doses of radiation, and any regulatory basis that presumes otherwise is unfounded.

But it is not just the LNT model that is problematic. LNT is used alongside two other concepts that exacerbate the problem significantly. The first is the precautionary principle, which is an approach to environmental policy that is built around taking preventive action in the face of uncertainty, shifting the burden of proof to the proponents of an activity, exploring a wide range of alternatives to possibly harmful actions, and increasing public participation in decision-making.[11]

While the NRC does not formally endorse the precautionary principle, its concepts are reflected in the commission's regulations. This is a significant problem when a preponderance of the NRC's regulations around protecting against low-dose radiation rely on the LNT model.

But there isn't enough uncertainty around low-dose radiation and LNT to justify using the precautionary principle. In fact, countless studies and decades of experience show that LNT is wrong. So shifting the burden of proof to the proponents makes no sense. There is nothing to prove. There is also no need to explore a wide range of alternatives to harmful actions because some exposure to low-dose radiation is demonstrably not harmful. Finally, increasing public participation in decision-making around an activity that is not harmful is pointless. We don't need the public to weigh in on perfectly safe decisions, and when we create a process to do that, it perpetuates the idea that something is harmful even though evidence proves otherwise. Further, it subjects decision-making to an arduous process that often leads to uncertainty and stasis. Essentially, this is just another tool used by the antinuclear movement to stop nuclear energy in its tracks.

This then relates to the second concept: as low as reasonably achievable, or ALARA. According to the NRC, in practice this means "making every reasonable effort to maintain exposures to ionizing radiation as far below the dose limits as practical," considering the state of technology, economic considerations, and impact on public health and safety.[12] ALARA is not just some loosely defined operational philosophy—it is codified in federal law in the NRC's Standards for Protection Against Radiation.[13]

If those dose limits are informed by an LNT model that presumes that any exposure to radiation increases the risk of cancer, then there is no clear limiting factor to calibrate the regulatory burden. In fact, this is exactly why the NRC uses ALARA. According to the agency, "ALARA is necessary because it is assumed that every radiation exposure carries with it some risk."[14] The same document acknowledges there is no limiting factor when it says that "since it [ALARA] is not a quantitative limit, it is subjective to a degree."

In monetary policy, fiat currencies are currencies that are not backed by anything real, like gold and silver, and instead derive value by virtue of a government saying they are valuable. It's not a leap to conclude that a substantial portion of the regulation governing the American nuclear industry is essentially fiat regulation. It is not backed by anything real, and its power is derived simply by virtue of government saying it's powerful.

It's time to wipe the slate clean.

But a reform effort will not come easily or without cost. It will necessarily upset the entire body of regulation that governs nuclear power, disrupt the consensus among some establishment institutions, and undermine one of the most potent tools used by the antinuclear movement. It could also create opposition among supporters of other energy sources that depend on the high costs of nuclear power to remain competitive. But when one of the most significant cost drivers for nuclear energy is removed, the reward will be immense for humanity.

To be clear, this doesn't mean that protecting workers and the public from dangerous levels of radiation should not be a driving force behind commercial nuclear power regulation. What it does mean is that regulation should be based on actual risk, and regulatory agencies should focus on high-level protection standards that are based on current scientific knowledge.

BROADEN STATE AUTHORITY TO REGULATE NUCLEAR POWER

So long as we are revolutionizing *how* the nation regulates nuclear power, why not also revolutionize *who* regulates nuclear power?

The failure of federal bureaucrats to reasonably regulate this industry raises the legitimate question of who else could take on some of this responsibility. The answer is simple: the states. States could be authorized to take a larger role in nuclear power plant regulation. The Atomic Energy Act of 1954 already allows states to regulate some nuclear materials under the Agreement State Program.[15] States that ink agreements with the NRC under the program are authorized to regulate byproduct materials like isotopes, source materials like uranium and thorium, and limited quantities of special nuclear materials like plutonium and enriched uranium. Under these agreements, states can license, inspect, and enforce safety regulations.[16]

It wasn't always this way. In the beginning of the nuclear age, the federal government held authority over all nuclear activities. It soon became clear, however, that for the technology to move forward, Washington had to give up its monopoly. In 1954, this gave rise to the Atomic Energy

Act (AEA), which among other things provided the statutory framework under which private firms could engage in commercial nuclear endeavors.

As the technology advanced, states began voicing concerns about being shut out of overseeing activities taking place within their own borders. This led, in 1959, to an amendment to the AEA to establish the regulatory framework for what is now the Agreement State Program. Section 274 b of the AEA lays out in specific detail what states can oversee, where the federal government retains its authority, and under what conditions states can take on the additional responsibilities.

Though their authority was limited in scope, states quickly began establishing agreements with the federal regulator. The first state, Kentucky, signed its agreement in 1962, and today thirty-nine states have similar agreements with the NRC.

That authorities early on opened nuclear technology up to commercialization and limited state oversight demonstrates that the American nuclear program was initially receptive to figuring out the best way to move the technology forward. That same openness to change in the name of progress is required today, and one way to achieve it is to amend the AEA once again to broaden states' authority to regulate all aspects of nuclear power, including the construction and operation of commercial nuclear plants. Under such an expansion, states could regulate existing nuclear power plants and future advanced reactors, and the NRC could focus on growing demand for the capabilities where it already has a comparative advantage. This would include things like certifying new light-water reactor designs, reactor life extensions, power uprates, and bringing back online reactors that have been shuttered.

We should acknowledge that an obstacle for the NRC in permitting advanced reactors is their lack of experience regarding those technologies. In essence, the NRC must build that expertise before it can regulate those reactors. But instead of diverting resources that could be better used to fulfill current demand for what the NRC already knows, it would make more sense to pass that responsibility to the states, which would be very capable of building the technical bases required to regulate these new technologies. Not all states would take advantage of this opportunity, but some would want to invest the resources to build their own regulatory framework. In fact, many states already have nuclear regulatory infrastructure, and this could be expanded. For those states that don't want to do that, the NRC would still be available to do what it does now.

How would this work? States would have to demonstrate to the NRC that they had the expertise, legal framework, and resources to effectively regulate commercial nuclear activities. It may be that only power plants using a reactor design certified by the NRC would be eligible for state oversight at first. Other limitations could be applied as well. The program could, for example, be implemented in stages, with a state first having to demonstrate an ability to provide oversight of existing plants before its authority could be expanded to provide oversight of constructing and operating new plants. Very small nuclear plants—say those under 100 megawatts—could be completely overseen by states at all stages of development, given their lower risk profile.

This authority should also include oversight of fuel cycle facilities, such as spent fuel storage and technical processing. Giving states authority over the operation of such facilities could simply happen through a different cooperative arrangement. The NRC would retain the authority to monitor and control any nuclear-proliferation-sensitive materials that result from those activities and also be available to support the states and provide technical assistance. It should also remain responsible for coordinating responses to any major nuclear accidents.

The point is that US utilities have been safely operating large light-water reactors for over fifty years, yet our approach to regulation still treats these reactors as scary new technology. But the fact is that nuclear energy is not new or scary, and it is time that we start regulating it as the understood, safe industrial activity that it is.

FOCUS THE NRC ON AREAS WHERE IT HAS A COMPETITIVE ADVANTAGE OR ESTABLISHED EXPERTISE

One of the benefits of transferring some of the regulatory burden away from the NRC is that it would free up resources for something that the commission already has decades of experience in doing, which is regulating light-water reactors.

Like all government agencies, the NRC has numerous responsibilities and limited resources to carry them out, so it must prioritize. It makes sense for it to focus on regulating the large light-water reactors that make up America's commercial fleet. This is the technology used in every one of America's existing commercial power reactors, as well as the underlying technology of the eight reactor designs that have been approved by the NRC through its design certification process. It's safe to assume that America's commercial nuclear fleet will remain a large light-water fleet for some time to come. That doesn't mean that new technologies aren't on the horizon. In fact, nearly all the talk about nuclear energy in recent years has revolved around advanced reactor designs and SMRs, which are predominantly non-large light-water reactors.

While new submissions are occurring on a fairly regular basis, the NRC is in some level of pre-application discussions with around a dozen applicants and is even reviewing a few applications to construct power and research reactors.[17] While this demonstrates that there is authentic private-sector interest in moving advanced reactors forward, the fact that so much work needs to be done before the NRC can even accept applications also demonstrates that the agency, at least to a degree, is not yet fully prepared to meet this growing demand.

To its credit, the NRC has in recent years engaged in multiple efforts with stakeholders to identify how it can better organize itself to meet the needs of advanced reactors and to improve the regulatory process to facilitate permitting. As a result, it has reorganized some of its offices to better respond to the growing demand for advanced reactors and created license review teams to support that effort. Perhaps most importantly, per congressional requirement, the NRC has begun to develop a new licensing pathway, known as Part 53, that is catered to the technological diversity of advanced reactors.[18]

Despite this progress, concern remains that NRC is unprepared to efficiently process advanced reactor regulatory activities, and that the process still presents a significant obstacle to the broad acceptance of advanced reactors in the US.[19] Recent setbacks within the advanced reactor sector demonstrate this ongoing problem.[20] Moreover, all this activity to prepare for potential future advanced designs takes valuable resources away from the NRC's ability to serve growing demand to support America's existing reactor fleet.

In 2019, the NRC extended the license for Florida's Turkey Point Nuclear Units 3 and 4 a second time, allowing these reactors to operate for a total of eighty years, which was unprecedented at the time.[21] Since then, fifteen more reactors have applied for a second twenty-year extension. While eighty-eight reactors have already received approval of their first twenty-year extension, most of America's reactors will need a relicensing extension during the 2030s.[22]

And then there are power uprates, which is when nuclear plants seek regulatory authority to expand the amount of power they produce. According to one industry study released in 2023, over 50 percent of the sites surveyed were at least considering a power uprate.[23] And that was before the likelihood of future electricity shortages became clearer over the past year—so if anything, the demand for power uprates will grow even more substantially than was thought.

On top of all the extensions and uprates, resources will have to be devoted to ensuring a smooth process for a growing number of projects to restart shuttered reactors or revive stalled efforts to build new reactors. A long-term power purchase agreement between Palisades Energy and Wolverine Power Cooperative is bringing Michigan's 800 MW Palisades nuclear plant, which was shut down in 2022, back online in 2025. The agreement includes a provision for up to two new SMRs, rated at 300 MW each, to be built at the Palisades site.[24]

In Pennsylvania, Microsoft just entered a twenty-year power agreement with Constellation that will result in Three Mile Island Unit 1 restarting production in 2028. The plant will sell 800 MW of its power directly to Microsoft's data center.[25] This is popular among Pennsylvanians—according to a statewide poll, registered voters favor restarting Unit 1 by a two-to-one margin.[26]

Microsoft isn't alone. Goldman Sachs estimates that energy demand from data centers will grow 160 percent by 2030.[27] In March 2024, Amazon Web Services purchased a site from Talen Energy, adjacent to its Susquehanna nuclear plant, for $650 million. According to the power agreement, Amazon will purchase between 480 and 960 MW of electricity to power its data centers from Talen's 2,228 MW stake in the Susquehanna plant.[28]

And there is the prospect of growing demand for new large light-water reactors. While small reactors elicit plenty of excitement, the reality is that they will not likely solve big electricity shortage problems. And while the cost escalations associated with the Vogtle experience are often cited as a reason not to build new large rectors, there is a more optimistic way to look at it. The fact is that the AP1000s built at Vogtle were new reactor designs, and applying the lessons learned from that project, as well as those in China, will likely result in cost savings moving forward. While it may be popular to suggest that the era of big nuclear reactors is over, growing demand for massive amounts of reliable clean energy could be a market that only big nuclear can satisfy. If this is the case, as was demonstrated in the late 2000s, a lot of firms could be coming to the NRC with large light-water reactor permit applications.

Putting aside critiques of how the NRC regulates, the agency simply doesn't have the resources to do everything that a growing nuclear industry requires. The US Government Accountability Office agrees. In a 2023 report, it identified issues with insufficient staffing and poor communications with applicants, and concluded that the NRC is not ready to take on the new responsibility of a growing, more diverse nuclear sector.[29] The NRC has blamed these shortcomings on a lack of resources.

A lack of resources can be fixed in one of two ways: more resources can be made available or existing resources can be reallocated to focus on higher priorities. Given the exorbitant costs already associated with engaging with the NRC and the nation's out-of-control debt, there is likely little to be gained by raising fees or adding to the taxpayer's burden. But if the commission was relieved of some of its responsibility for overseeing every detail of well-understood technologies or having to develop new expertise that could be handled by the states, there would be more money to dedicate to the things it does best, which are also the things the industry needs right now. This prioritization should be clarified by both executive action and statute as necessary.

Additional resources could also be freed by taking better advantage of private-sector contractors. One way to do this would be for the NRC to expand efforts to certify private firms to provide technical and legal reviews of applications for design certifications, construction applications, and environmental work. This would free up resources within the NRC to be applied to things that only the commission can do and could even save money if the private firms were able provide the same level of service at lower cost.

The NRC has already suggested that it is open to relying more heavily on contractors to help make up for staffing shortfalls. The commission also acknowledged that one of its challenges in developing the workforce necessary to support advanced reactor regulation is that private firms are outcompeting them for talent. If the same people the NRC wants to hire are already working in private firms, why can't those firms be used to supplement the NRC's regulatory processes through contractual agreements? Rather than fight the market, the NRC should use it to its advantage.

A host of other reforms have been proposed to fix the inefficiency of the regulatory process for advanced reactors. In 2023, for example, the Idaho National Laboratory released a study that provided a comprehensive set of recommendations to streamline the NRC hearing process, expedite safety and environmental reviews, and improve the overall process for regulating advanced reactors.[30] The GAO study referenced above also provides several helpful recommendations. On their own, these recommendations will not bring about the sort of systemic change that is necessary, but if they're combined with the more ambitious reforms proposed here, meaningful change just might result.

DEVELOP REGULATORY ALTERNATIVES AROUND PRIVATE LIABILITY INSURANCE REQUIREMENTS

In the early days of American nuclear power, one of the obstacles to moving the technology out of the labs and into the marketplace was concern about who would be liable in the event of an accident at a privately owned nuclear facility. At the time, there were too many unknowns about commercial nuclear technology, which made establishing an affordable private liability insurance industry against nuclear accidents all but impossible. To address this problem, Congress amended the Atomic Energy Act in 1957 with what is known as the Price-Anderson Act. This act established a system of liability that on the one hand caps exposure to NRC licensees (i.e., nuclear power plant operators) and on the other ensures that adequate funds are available to pay for damages that may result from a nuclear accident.

Put simply, the system consists of two tiers that set the requirements for coverage and the limits of liability. The first tier requires that every power plant owner acquire private insurance coverage in the amount of $500 million per plant. The second tier, which kicks in when damages exceed $500 million, is an industry-wide self-funded insurance arrangement where each reactor licensee is responsible for an equal portion of the damages that result from a nuclear accident at any covered plant. The liability for the second tier is capped at $158 million per reactor. A 5 percent surcharge can also be imposed on top of the $158 million, should damages exceed available funds, bringing the total potential second-tier compensation available to $165.9 million per reactor.[31] Since there are ninety-four reactors operating in the United States, that means there is a total of $16.1 billion, when first-tier compensation is included, available to pay out for damages in the case of an accident.

For better or worse, the public policy trade-off is that the public knows there is $16.1 billion in non-taxpayer funds readily available to pay for any damages that result from an accident and the nuclear industry gets a cap on its potential liability. Critics argue that capping liability is little more than a subsidy that hides the true risks of nuclear power and socializes the costs for any accidents that cause damage that exceeds those amounts. It allows insurance premiums to reflect capped liability, rather than, opponents argue, the true risks of nuclear energy. Proponents, on the other hand, contend that this system creates clear lines of responsibility and requires that ample, easily accessible funds are available should an accident occur.

The truth probably lies somewhere in the middle.

But it's important to acknowledge that there was a valid justification for Price-Anderson protection in the early days of nuclear power. Putting aside the national interest pursuits that early officials had in helping move nuclear into commercialization, the fact is that commercializing nuclear power came with a lot of risk and unknowns. But that's no longer the case, certainly not with large light-water reactors. Since the act's implementation in 1957, fewer than 250 claims have been filed, for a total payout of $522 million.[32] Of this, $71 million was paid to address claims from America's largest nuclear accident, Three Mile Island. In other words, in nearly seventy years, the total in paid claims barely exceeds the first tier of coverage held by each individual reactor.

Whether this shows that Price-Anderson is no longer needed is a debate better left for another time. What it does show is that the public health risks posed by commercial reactors fall well below liability coverage requirements. Nonetheless, so long as Price-Anderson exists and the NRC indemnifies licensees against damages above a certain amount, it is understandable (if not preferable) that the commission applies the heavy regulatory hand it does.

But what if licensees were *not* indemnified by the NRC? That would shift the full burden of risk away from the taxpayer and onto the private sector, without any reduction in safety standards. Regulatory agencies would still set the standards with which reactor operators must comply, but the regulatory burden would be significantly decreased. In fact, it's possible that such a system would ensure even greater safety as private firms would need to take on full responsibility for their actions and would therefore be even more incentivized to ensure safe operations. Greater levels of safety would also result in lower insurance premiums. This whole alternative framework would give firms the time and ability to focus on actual improvements to safety rather than dealing with federal bureaucracy.

Critics of this proposed approach might argue that while a greatly decreased regulatory burden is tempting, no firm would actually forgo Price-Anderson protection and no private insurer would provide full liability coverage absent the caps provided by Price-Anderson. The firm that currently supplies nuclear liability insurance to all licensees within the existing system argues that the entire sector could become uninsurable without Price-Anderson.[33]

Maybe that's true. Maybe it's not.

In fact, one representative from another major insurance firm sees it differently.[34] In 2014, a managing director of the power, nuclear, and construction division at Price Forbes, said, "It is my belief that insurers could do more. We could provide cost-effective, materially higher financial support for the nuclear industry, reducing the burden of accident costs that currently falls to governments and taxpayers." He added that "transferring the nuclear accident risk at its true scale to the insurance market is entirely feasible and well within the financial capability of the global market; it just needs some new thinking to enable this capacity to be utilized."

The basis of his argument is that the insurance industry, like the nuclear industry, has matured and become more sophisticated since Price-Anderson went into effect. He points out that today's insurers are fully capable of paying out massive claims, and he points to the $72 billion paid out after Hurricane Katrina as evidence.

When progress in the insurance industry is combined with the promise of enhanced safety offered by advanced reactors, it's not hard to imagine private insurers indeed covering nuclear liability outside of Price-Anderson. Consider, too, that some of the world's most successful entrepreneurs are actively pursuing advanced reactors. Perhaps they are both confident enough in the technology and disgruntled enough with the current system of regulation that they would want to take advantage of such an alternative framework.

Even if private insurance firms were not willing to take on the full coverage requirements set by state and/or federal regulators, a well-funded entrepreneur could make up the difference. Or perhaps advanced reactor firms could pool their resources to self-insure. The point is that no one really knows what the market might come up with if given the opportunity, and when combined with modernized low-dose radiation standards, such a reform could truly revolutionize how nuclear power plants are built and financed in the United States.

While such a system could be designed in any number of ways, it would undoubtedly have certain key elements. First, the regulators would have to create new binding guidance that defines the safety and security criteria for plants using the new process. Regulators would also have to set liability coverage requirements for licensees, define exactly what must be covered, and clarify what constitutes coverage. These requirements could be made part of the design certification process, and firms could be required to use a certified design to take advantage of the new process.

The point here is not to abolish Price-Anderson—at least not yet. Rather, it's to open the market up to new and innovative ways to finance, build, and operate safe reactors. This approach also acknowledges that the NRC is not the font of all nuclear safety. The nuclear industry has a deep and proven culture of safety that informs its decision-making. It has created private-sector organizations that promote and enforce that culture.

None of this is to say that the NRC deserves no credit for the nuclear industry's stellar safety record, but it does argue that if we want a revolution in commercial nuclear energy, we need to think outside of how things have been done for the past seventy years—and that includes how the industry is insured.

APPLY MARKET PRINCIPLES TO SPENT NUCLEAR FUEL POLICY REFORM

The current approach to managing used nuclear fuel is systemically broken. It was developed to support a nuclear industry that was largely believed to be in decline. That is no longer the case. When the federal government promised to take responsibility for used fuel and safely dispose of it, this removed any incentive the private sector may have had to develop better ways to manage fuel that might have been more consistent with an emerging nuclear industry. To make matters worse, the federal government has proven incapable of fulfilling its obligations to dispose of the fuel.

The current system is driven by government preoccupations and politics. There is little connection between federal used-fuel management programs and the real-world needs of the nuclear industry. Any successful plan must grow out of the private sector. The time has come for the federal government to step aside and allow utilities, nuclear technology companies, and consumers to manage used nuclear fuel.

Overhauling the nation's nuclear-waste management regime will not be easy. Among other things, it will require a significant amendment of the Nuclear Waste Policy Act and a long-term commitment by Congress, the administration, and the industry itself. But developing such a system would put the United States well on its way to reestablishing itself as a global leader in nuclear energy.

FIX NUCLEAR FUEL MARKETS

America's nuclear fuel markets are broken.

Our reactors are powered largely by uranium, and although uranium is an abundant mineral around the world, few countries have the ability to enrich it for use in nuclear power plants. Russia controls around 46 percent of global enrichment capacity; America controls only 9.5 percent.[35] This is an energy security problem because the United States is the world's largest consumer of LEU but is able to produce only around 20 percent of its requirements domestically. The rest comes from enrichment facilities in the United Kingdom, Germany, and the Netherlands. Until recently, around 25 percent came from Russia.[36]

But this wasn't always the case. America was largely self-sufficient in uranium and LEU production until the end of the Cold War, when capacity in both began to decline.[37] America went from self-sufficiency to extreme dependency in a relatively short period of time—and too much of that dependency was on Russia.

No one cared much about this until 2022, when Russia invaded Ukraine and America's reliance on Russian uranium became a real problem.[38] Not only was the US dependent on Russia for uranium and related nuclear fuel services, but roughly a billion dollars was flowing to Russian state-owned enterprises annually as a result.[39]

Promptly after the invasion, President Biden stopped oil imports from Russia. More recently, Congress and the administration have acted to ban uranium imports. The Prohibiting Russian Uranium Imports Act, which lasts until 2040, bans the import of LEU from Russia or any Russian entity; prevents black market imports by banning LEU that "is determined to have been exchanged with, swapped for, or otherwise obtained" in an effort to circumvent the ban; and creates a waiver process where the ban can be lifted on certain imports for a limited amount of time.

A long-term ban is critical because expanding enrichment is time-consuming and expensive. Investors won't expand capacity to make up for Russian supply if the ban may be terminated or waived, which would make low-cost Russian LEU once again available to US buyers. A long-term ban on Russian imports is necessary to provide the market certainty required to justify investment in a broader expansion in the sector.[40]

But there are two potential stumbling blocks to expanding US nuclear fuel capacity. The first is abuse of the waiver process that the act established to allow some Russian imports through 2027 under certain conditions. While a waiver process may be necessary in specific cases, waivers must not be granted unless there is no alternative source for the LEU. The second obstacle to expanding domestic nuclear fuel production capabilities as expeditiously as possible is government. The regulatory process to expand, build, and operate new domestic LEU capacity has to be streamlined. Government needs to get out of the way.

The private sector has been safely building and operating commercial enrichment facilities around the world for decades, and it should be allowed to expand US capacity with as little interference from Washington as possible. Furthermore, domestic uranium mining has almost disappeared. Opening new US mines is extremely difficult, but a robust domestic mining industry helps to protect against foreign supply disruptions.[41] Current policies mean domestic uranium miners miss out on the opportunity to provide secure supplies to American reactors, and reactor operators must continue their dependence on foreign suppliers.

Further, policies should be set in place that allow for a more seamless diversification of American nuclear fuel supply to include alternative fuel cycles like those using thorium or spent nuclear fuel.

The time has come to move to finally fix America's nuclear fuel industry by freeing it to once again the lead the world into a clean, abundant, and safe nuclear future.

ACKNOWLEDGMENTS

As is the case with any major endeavor, the list of people without whom the accomplishment could not have occurred is larger than the space available—and such is the circumstance of this book. Nonetheless, there are people to whom I am eternally grateful and without whom this book would not exist … so blame them.

I will start by thanking my intrepid boss, Diana Furchtgott-Roth. She not only gave me the space within my day job to write this book and worked tirelessly to help get the word out that it was coming, but also pushed me to take on the project in the first place. You are a great boss, Diana!

Then there is my partner in this crime and a handful of others, Andrew Weiss. Andrew worked with me from day one on this book, providing research support, tracking down citations, reviewing the writing, and taking on countless other thankless tasks. He also listens to my stories and laughs at my jokes, which counts for a lot. Thank you, Andrew.

I also want to thank Steve Moore. Steve has been a trusted colleague and friend for many years, and that he agreed to provide the foreword to this book is humbling, to say the least.

I would also like to thank Dean Baxendale and Optimum Publishing for supporting me on this project. You have been a pleasure to work with, and I sincerely appreciate everything you have done. And of course, this book would be the literary version of a cheap Jackson Pollack imitation—that is, a bunch of blotches on the wall—if not for the skilled editing of

Janice Weaver. As anyone who writes for a living knows, it's all about the editor!

Not to be dramatic, and certainly not to transfer my sins to others, but this book would not exist if not for two very good longtime friends and former colleagues, Nick Loris and Katie Tubb. I haven't worked with Nick and Katie for a while, but they were both integral to the development of my approach to nuclear energy policy and to much of my previous work on the issue, which is largely reflected in this book. More important than that, however, is that they have inspired me, made me laugh, and made me smarter, and they are two of the only people outside of the ones I'm about to mention who can put me in my place. So for all of that and more, thank you both.

And then there is my wife, Terri, and my daughter, Madeline. I'm so lucky to have you both in my life. Thank you for everything you do for me and for our family.

Finally, I want to thank my colleagues, past and present, at the Heritage Foundation. At Heritage, we talk about making the American dream accessible to everyone, and the policies we work on each day bring us closer to that reality. But you also do that on an individual basis. At least you did that for me. I still don't understand why, but you've allowed me to take on responsibilities that I've had no business taking on. You've allowed me to have meaningful conversations with people whose stories will be told in history books. You've allowed me to make mistakes and grow. And you did all this while allowing me to be me. You could fire me tomorrow, and I'd still be forever grateful. Thank you.

ENDNOTES

CHAPTER 1: WHY NUCLEAR ENERGY?

1. Epstein, Alex, *Fossil Future: Why Global Human Flourishing Requires More Oil, Coal, and Natural Gas—Not Less*, New York: Portfolio/Penguin, 2022, p. 23.

2. Hodges, Paul, "Rising Life Expectancy Enabled Industrial Revolution to Occur," *Chemicals and the Economy* (blog), February 27, 2015. https://www.icis.com/chemicals-and-the-economy/2015/02/rising-life-expectancy-enabled-industrial-revolution-to-occur/.

3. Hodges, "Rising Life Expectancy."

4. O'Neil, Aaron, "Gross Domestic Product (GDP) Per Capita in Selected Global Regions in 2019," Statista, February 2, 2024, https://www.statista.com/statistics/256413/gross-domestic-product-per-capita-in-selected-global-regions/.

5. Dayaratna, Kevin, Diana Furchtgott-Roth, Miles Pollard, and Richard Stern, "Powering Human Advancement: Why the World Needs Affordable and Reliable Energy," Heritage Foundation, December 14, 2023, https://www.heritage.org/energy/report/powering-human-advancement-why-the-world-needs-affordable-and-reliable-energy.

6. "Gross Domestic Product, Fourth Quarter and Year 2022 (Third Estimate), GDP by Industry, and Corporate Profits," US Bureau of Economic Analysis (BEA), March 30, 2023, https://www.bea.gov/news/2023/gross-domestic-product-fourth-quarter-and-year-2022-third-estimate-gdp-industry-and.

7. "Our Nation's Air: Trends through 2021," US Environmental Protection Agency (EPA), n.d. https://gispub.epa.gov/air/trendsreport/2022/#home.

8. Shellenberger, Michael, "Why Disasters Have Declined," *Forbes*, January 11, 2022, https://www.forbes.com/sites/michaelshellenberger/2022/01/10/why-disasters-have-declined/.

9. Soon, Willie, "Ph.D. Scientist Willie Soon Easily Debunks Climate Change Propaganda," Heartland Institute, March 9, 2023, YouTube, https://www.youtube.com/watch?v=_Ma4aSFlF_Q.

10. Dayaratna, Kevin D., "The Unsustainable Costs of President Biden's Climate Agenda," Heritage Foundation, June 16, 2022, https://www.heritage.org/energy-economics/report/the-unsustainable-costs-president-bidens-climate-agenda.

11. The OECD countries as of 2024 are Australia, Austria, Belgium, Canada, Chile, Colombia, Costa Rica, Czechia, Denmark, Estonia, Finland, France, Germany, Greece, Hungary, Iceland, Ireland, Israel, Italy, Japan, Korea, Latvia, Lithuania, Luxembourg, Mexico, Netherlands, New Zealand, Norway, Poland, Portugal, Slovak Republic, Slovenia, Spain, Sweden, Switzerland, Türkiye, United Kingdom, United States.

12. "U.S. Energy Consumption Increases Between 0% and 15% by 2050," US Energy Information Administration (EIA), April 3, 2023, https://www.eia.gov/todayinenergy/detail.php?id=56040.

13. "2023 Long-Term Reliability Assessment," North American Electric Reliability Corporation (NERC), December 2023, https://www.nerc.com/pa/RAPA/ra/Reliability%20Assessments%20DL/NERC_LTRA_2023.pdf.

14. "EIA Projections Indicate Global Energy Consumption Increases through 2050, Outpacing Efficiency Gains and Driving Continued Emissions Growth," US EIA, press release, October 11, 2023. https://www.eia.gov/pressroom/releases/press542.php.

15. Department of Energy Organization Act, Pub. L. No. 95-91 (1977), https://www.congress.gov/bill/95th-congress/senate-bill/826/text.

16. "Oil and Gas Supply Module," EIA, March 2024, https://www.eia.gov/outlooks/aeo/assumptions/pdf/oilgas.pdf.

17. "2024 North American Energy Inventory," Institute for Energy Research (IER), May 14, 2024, https://www.instituteforenergyresearch.org/wp-content/uploads/2024/05/2024-North-American-Energy-Inventory.pdf.

18. "2024 North American Energy Inventory," IER.

19. "U.S. Energy Facts Explained: Imports and Exports," EIA, July 15, 2024, https://www.eia.gov/energyexplained/us-energy-facts/imports-and-exports.php.

20. Wells, B.A., and K.L. Wells, "Shooters: A 'Fracking' History," American Oil & Gas Historical Society, September 1, 2007, https://aoghs.org/technology/hydraulic-fracturing/.

21. "German Industry to Pay 40% More for Energy Than Pre-Crisis—Study Says," Reuters, January 30, 2023, https://www.reuters.com/business/energy/german-industry-pay-40-more-energy-than-pre-crisis-study-says-2023-01-30/.

22. "Business Expectations of German Companies Drop Massively," German Chamber of Commerce and Industry (DIHK), May 24, 2022, https://www.dihk.de/en/german-economy/business-expectations-of-german-companies-drop-massively-72104.

23. Alkousaa, Riham, "German Economy Minister Under Fire as German Companies Sound Alarm on Energy Prices," Reuters, September 7, 2022, https://www.reuters.com/business/energy/german-economy-minister-under-fire-german-companies-sound-alarm-energy-prices-2022-09-07/.

24. "Germany: Countries & Regions," International Energy Agency (IEA), 2022, https://www.iea.org/countries/germany.

25. "Statistical Review of World Energy, 2021," British Petroleum (BP), https://www.bp.com/content/dam/bp/business-sites/en/global/corporate/pdfs/energy-economics/statistical-review/bp-stats-review-2021-coal.pdf.

26. "Coal Production and Consumption See Rebound in 2021," Eurostat, May 2, 2022. https://ec.europa.eu/eurostat/web/products-eurostat-news/-/ddn-20220502-2#.

27. "European Gas Reserves," International Association of Oil & Gas Producers (IOGP) Europe, n.d., https://iogpeurope.org/european-gas-reserves/.

28. "Natural Gas Consumption in the European Union from 1998 to 2023," Statista, August 26, 2023, https://www.statista.com/statistics/265406/natural-gas-consumption-in-the-eu-in-cubic-meters/.

29. Kimani, Alex, "Why Europe Won't Exploit Its Huge Gas Reserves," Oilprice.com, September 11, 2022, https://oilprice.com/Energy/Energy-General/Why-Europe-Wont-Exploit-Its-Huge-Gas-Reserves.html.

30. "Nuclear Power in the European Union," World Nuclear Association, August 13, 2024, https://world-nuclear.org/information-library/country-profiles/others/european-union.

31. Alkousaa, Riham, and Christian Kraemer, "Germany Set to Miss Net Zero by 2045 Target as Climate Efforts Falter," Reuters,

August 22, 2023, https://www.reuters.com/business/environment/germanys-climate-efforts-not-enough-hit-2030-targets-experts-say-2023-08-22/.

32. Sgaravatti, Giovanni, Simone Tagliapietra, and Georg Zachmann, "National Policies to Shield Consumers from Rising Energy Prices," Bruegel, October 21, 2022, https://www.bruegel.org/dataset/national-policies-shield-consumers-rising-energy-prices.

33. Kennedy, Brian, Cary Funk, and Alec Tyson, "Majorities of Americans Say Too Little Is Being Done on Key Areas of Environmental Protection," Pew Research Center Science & Society, June 28, 2023, https://www.pewresearch.org/science/2023/06/28/3-majorities-of-americans-say-too-little-is-being-done-on-key-areas-of-environmental-protection/.

34. "What Is U.S. Electricity Generation by Energy Source?" EIA, February 29, 2024, https://www.eia.gov/tools/faqs/faq.php?id=427&t=3.

35. "Nuclear Power in the World Today," World Nuclear Association, March 2024, https://world-nuclear.org/information-library/current-and-future-generation/nuclear-power-in-the-world-today.aspx.

36. Ritchie, Hannah, "How Does the Land Use of Different Electricity Sources Compare?" Our World in Data, June 16, 2022, https:// ourworldindata.org/land-use-per-energy-source.

37. "Electricity Data," EIA, 2016, https://www.eia.gov/electricity/monthly/epm_table_grapher.php?t=epmt_6_07_b.

38. Mathew, Manuel S., Surya Teja Kandukuri, and Christian W. Omlin, "Estimation of Wind Turbine Performance Degradation with Deep Neural Networks," *PHM Society European Conference* 7, no. 1 (June 29, 2022): 351–59, https://doi.org/10.36001/phme.2022.v7i1.3328.

39. "Radioactive Effluent and Environmental Reports," US Nuclear Regulatory Commission (NRC), June 15, 2023, https://www.nrc.gov/reactors/operating/ops-experience/tritium/plant-info.html.

40. Shellenberger, Michael, "It Sounds Crazy, but Fukushima, Chernobyl, and Three Mile Island Show Why Nuclear Is Inherently Safe," *Forbes*, March 11, 2019, https://www.forbes.com/sites/michaelshellenberger/2019/03/11/it-sounds-crazy-but-fukushima-chernobyl-and-three-mile-island-show-why-nuclear-is-inherently-safe/?sh=1a1d6c6c1688.

41. "Nuclear Waste Disposal," US Government Accountability Office (GAO), https://www.gao.gov/nuclear-waste-disposal.

42. Wheeler, Andrew R., "The Need to Examine the Life Cycles of All Energy Sources: A Closer Look at Renewable-Energy Disposal," Heritage Foundation, September 20, 2021, https://www.heritage.org/renewable-energy/report/the-need-examine-the-life-cycles-all-energy-sources-closer-look-renewable.

43. Gil, Laura, "Finland's Spent Fuel Repository a 'Game Changer' for the Nuclear Industry, Director General Grossi Says," International Atomic Energy Agency (IAEA), November 26, 2020, https://www.iaea.org/newscenter/news/finlands-spent-fuel-repository-a-game-changer-for-the-nuclear-industry-director-general-grossi-says.

44. "What Is U.S. Electricity Generation by Energy Source?" EIA.

45. "Shippingport Nuclear Power Station," American Society of Mechanical Engineers, 2018, https://www.asme.org/about-asme/engineering-history/landmarks/47-shippingport-nuclear-power-station.

46. "Nuclear Power in the USA," World Nuclear Association, March 2024, https://world-nuclear.org/information-library/country-profiles/countries-t-z/usa-nuclear-power.

47. "China Continues Rapid Growth of Nuclear Power Capacity," EIA, May 6, 2024, https://www.eia.gov/todayinenergy/detail.php?id=61927#.

48. "Nuclear Power in the USA," World Nuclear Association.

49. "At COP28, Countries Launch Declaration to Triple Nuclear Energy Capacity by 2050, Recognizing the Key Role of Nuclear Energy in Reaching Net Zero," Department of Energy, December 1, 2023, https://www.energy.gov/articles/cop28-countries-launch-declaration-triple-nuclear-energy-capacity-2050-recognizing-key.

50. Spencer, Jack. "Nuclear Waste Policy Amendments Act of 2008: Modernizing Spent Fuel Management in the U.S.," Heritage Foundation, March 6, 2008, https://www.heritage.org/environment/report/nuclear-waste-policy-amendments-act-2008-modernizing-spent-fuelmanagement-the-us.

CHAPTER 2: SECURITY THROUGH AFFORDABLE, CLEAN, SAFE ENERGY

1. Lovering, Jessica R., Arthur Yip, and Ted Nordhaus, "Historical Construction Costs of Global Nuclear Power Reactors," *Energy Policy* 91 (April 2016): 371–82, https://doi.org/10.1016/j.enpol.2016.01.011.

2. Ritchie, "How Does the Land Use of Different Electricity Sources Compare?"

3. "Economics of Nuclear Power," World Nuclear Association, August 2022, https://world-nuclear.org/information-library/economic-aspects/economics-of-nuclear-power.aspx.

4. "Economics of Nuclear Power," World Nuclear Association.

5. Nuclear Energy Agency (NEA), *Projected Costs of Generating Electricity—2020 Edition* (Paris: OECD Publishing, 2020), https://www.oecd-nea.org/jcms/pl_51110/projected-costs-of-generating-electricity-2020-edition?details=true.

6. "Electricity Explained: Prices and Factors Affecting Prices," EIA, 2016, https://www.eia.gov/energyexplained/electricity/prices-and-factors-affecting-prices.php.

7. Durand, Patty, "Plant Vogtle: Not a Star, but a Tragedy for the People of Georgia," *POWER* magazine, August 11, 2023, https://www.powermag.com/blog/plant-vogtle-not-a-star-but-a-tragedy-for-the-people-of-georgia/.

8. "Producer Price Index by Commodity: Special Indexes: Construction Materials," FRED, Federal Reserve Bank of St. Louis, January 1, 1947, https://fred.stlouisfed.org/series/WPUSI012011.

9. "Nuclear Costs in Context," Nuclear Energy Institute (NEI), November 2021, https://nei.org/CorporateSite/media/filefolder/resources/reports-and-briefs/Nuclear-Costs-in-Context-2021.pdf. The levelized cost of electricity (LCOE) is calculated based on an installed capacity of 1,117 MW with a 91 percent capacity factor, an operating cost of $26.86 per MWh, a sixty-year lifespan, and a 7 percent cost of capital.

10. "Electric Power Monthly," EIA, 2024, https://www.eia.gov/electricity/monthly/epm_table_grapher.php?t=epmt_5_6_a.

11. "2023 H 5549 State of Rhode Island," 2023, https://webserver.rilegislature.gov/BillText/BillText23/HouseText23/H5549.pdf.

12. "Fact Sheet: Biden–Harris Administration Announces New Steps to Bolster Domestic Nuclear Industry and Advance America's Clean Energy Future," White House, May 29, 2024, https://www.whitehouse.gov/briefing-room/statements-releases/2024/05/29/fact-sheet-biden-harris-administration-announces-new-steps-to-bolster-domestic-nuclear-industry-and-advance-americas-clean-energy-future/.

13. Melillo, Jerry M., Terese Richmond, and Gary W. Yohe, eds., *Climate Change Impacts in the United States: The Third National Climate Assessment*, US Global Change Research Program, 2014, https://nca2014.globalchange.gov/report.
14. Ferrante, Lucas, and Philip M. Fearnside, "The Amazon: Biofuels Plan Will Drive Deforestation," *Nature* 577, 170 (2020), https://doi.org/10.1038/d41586-020-00005-8.
15. Van Veen, Kelsi, and Alex Melton, "Rare Earth Elements Supply Chains, Part 1: An Update on Global Production and Trade," US International Trade Commission, December 2020, https://www.usitc.gov/publications/332/executive_briefings/ebot_rare_earths_part_1.pdf.
16. Zettelmeyer, Jeromin, Simone Tagliapietra, Georg Zachmann, and Conall Heussaff, "Beating the European Energy Crisis," International Monetary Fund, December 2022, https://www.imf.org/en/Publications/fandd/issues/2022/12/beating-the-european-energy-crisis-Zettelmeyer.
17. "Nuclear Power in Russia," World Nuclear Association, September 7, 2024, https://world-nuclear.org/information-library/country-profiles/countries-o-s/russia-nuclear-power.
18. "Nuclear Power in China," World Nuclear Association, August 13, 2024, https://world-nuclear.org/information-library/country-profiles/countries-a-f/china-nuclear-power.
19. "China Is Building Nuclear Reactors Faster Than Any Other Country," *Economist*. November 30, 2023, https://www.economist.com/china/2023/11/30/china-is-building-nuclear-reactors-faster-than-any-other-country.

CHAPTER 3: PLAGUED BY MYTHS

1. Spencer, Jack, and Nicolas Loris, "Dispelling Myths About Nuclear Energy," Heritage Foundation, December 3, 2007, https://www.heritage.org/environment/report/dispelling-myths-about-nuclear-energy.
2. "Radiation Sources and Doses," EPA, March 28, 2019, https://www.epa.gov/radiation/radiation-sources-and-doses.
3. "Radiation Sources and Doses," EPA.
4. Mettler, Fred A., and Arthur C. Upton, *Medical Effects of Ionizing Radiation*,

3rd ed. (Philadelphia: Elsevier, 2008), https://www.sciencedirect.com/topics/medicine-and-dentistry/linear-no-threshold-model.

5. Cuttler, Jerry M., "The LNT Issue Is about Politics and Economics, Not Safety," *Dose-Response* 18, 3 (2020), https://doi.org/10.1177/1559325820949066. See also Selby, P. B., "The Selby-Russell Dispute Regarding the Nonreporting of Critical Data in the Mega-Mouse Experiments of Drs William and Liane Russell That Spanned Many Decades: What Happened, Current Status, and Some Ramifications," *Dose-Response* 18, 1 (2020), https://doi.org/10.1177/1559325819900714.

6. "Ukraine: Current Status of Nuclear Power Installations," NEA, June 4, 2024, https://www.oecd-nea.org/jcms/pl_66130/ukraine-current-status-of-nuclear-power-installations.

7. "NRC: Emergency Preparedness in Response to Terrorism," NRC, August 28, 2024, https://www.nrc.gov/about-nrc/emerg-preparedness/about-emerg-preparedness/response-terrorism.html.

8. Connolly, Kevin, "A Historical Review of the Safe Transport of Spent Nuclear Fuel," Department of Energy (DOE), August 31, 2016, https://www.energy.gov/ne/articles/historical-review-safe-transport-spent-nuclear-fuel.

9. "NRC: Physical Protection," NRC, March 11, 2020, https://www.nrc.gov/security/domestic/phys-protect.html.

10. "5 Fast Facts About Spent Nuclear Fuel," DOE, October 3, 2022, https://www.energy.gov/ne/articles/5-fast-facts-about-spent-nuclear-fuel.

11. "Radiation: Health Consequences of the Fukushima Nuclear Accident," World Health Organization (WHO), March 10, 2016, https://www.who.int/news-room/questions-and-answers/item/health-consequences-of-fukushima-nuclear-accident.

12. "Chernobyl: The True Scale of the Accident," WHO, September 5, 2005, https://www.who.int/news/item/05-09-2005-chernobyl-the-true-scale-of-the-accident.

13. "Chernobyl: The True Scale of the Accident," WHO.

14. "Radiation Sources and Doses," EPA.

15. Britannica, "Fukushima Accident," *Encyclopædia Britannica*, September 10, 2024, https://www.britannica.com/event/Fukushima-accident.

16. "Davis-Besse Reactor Pressure Vessel Head Degradation,"

NRC, August 2008, https://international.anl.gov/training/materials%5CA8%5CMoiseytseva%5CbrO353r1.pdf.

17. "NRC Sends Specialists to Vermont Yankee to Review Cooling Tower Leak," NRC, July 13, 2008, https://www.nrc.gov/docs/ML0819/ML081960039.pdf.

18. "North Anna Nuclear Power Plant Seismic Event Plant Seismic Event," NRC, August 30, 2011, https://www.nrc.gov/docs/ML1124/ML112420551.pdf.

CHAPTER 4: THREE MILE ISLAND, CHERNOBYL, AND FUKUSHIMA DAIICHI

1. "Health Studies Find No Cancer Link to TMI," American Nuclear Society, July 11, 2012, https://wx1.ans.org/pi/resources/sptopics/tmi/healthstudies.php.

2. "Combined License Holders for New Reactors," NRC, July 31, 2023, https://www.nrc.gov/reactors/new-reactors/large-lwr/col-holder.html.

3. "Backgrounder on the Three Mile Island Accident," NRC, March 28, 2024, https://www.nrc.gov/reading-rm/doc-collections/fact-sheets/3mile-isle.html.

4. "Chernobyl: The True Scale of the Accident," WHO, September 5, 2005, https://www.who.int/news-room/detail/05-09-2005-chernobyl-the-true-scale-of-the-accident.

5. "Chernobyl: The True Scale of the Accident," WHO.

6. "Chernobyl Accident 1986," World Nuclear Association, April 26, 2024, https://world-nuclear.org/information-library/safety-and-security/safety-of-plants/chernobyl-accident.

7. "The Chernobyl Accident," United Nations Scientific Committee on the Effects of Atomic Radiation, 2008, https://www.unscear.org/unscear/en/areas-of-work/chernobyl.html.

8. "Chernobyl: The True Scale of the Accident," WHO.

9. "On This Day: 2011 Tohoku Earthquake and Tsunami," National Centers for Environmental Information (NCEI), March 11, 2021, https://www.ncei.noaa.gov/news/day-2011-japan-earthquake-and-tsunami.

10. "Backgrounder on NRC Response to Lessons Learned from Fukushima," NRC, October 18, 2022, https://www.nrc.gov/reading-rm/doc-collections/fact-sheets/japan-events.html.

11. "Fukushima Daiichi Accident," World Nuclear Association, April 29, 2024, https://world-nuclear.org/information-library/safety-and-security/safety-of-plants/fukushima-daiichi-accident.

12. "Japan Confirms First Fukushima Worker Death from Radiation," *BBC News*, September 5, 2018, https://www.bbc.com/news/world-asia-45423575.

13. Suzuki, Yuriko, Hirooki Yabe, Seiji Yasumura, Tetsuya Ohira, Shin-Ichi Niwa, Akira Ohtsuru, Hirobumi Mashiko, Masaharu Maeda, and Masafumi Abe, "Psychological Distress and the Perception of Radiation Risks: The Fukushima Health Management Survey," *Bulletin of the World Health Organization* 93, no. 9 (June 15, 2015): 598–605, https://doi.org/10.2471/blt.14.146498.

14. "Nuclear Power Plants and Earthquakes," World Nuclear Association, March 5, 2021, https://world-nuclear.org/information-library/safety-and-security/safety-of-plants/nuclear-power-plants-and-earthquakes.

15. "Diverse and Flexible Coping Strategies (FLEX) Implementation Guide," NRC, December 2016, https://www.nrc.gov/docs/ML1635/ML16354B421.pdf.

16. "The Nexus Between Safety and Operational Performance in the U.S. Nuclear Industry," Nuclear Energy Institute (NEI), March 2020, https://nei.org/CorporateSite/media/filefolder/resources/reports-and-briefs/NEI-20-04-The-Nexus-Between-Safety-and-Operational-Performance-in-the-US-Nuclear-Industry.pdf.

CHAPTER 5: THE BROKEN POWER PLANT FALLACY

1. Tanner, Jari, "Europe's Most Powerful Nuclear Reactor Kicks off in Finland," *AP News*, April 16, 2023, https://apnews.com/article/finland-energy-nuclear-power-reactor-741341cfdf79e655a2a680e1b1130917.

2. "Another Delay, Cost Bump, for Flamanville-3," *Nuclear Newswire*, January 13, 2022, https://www.ans.org/news/article-3573/another-delay-cost-bump-for-flamanville3/. See also "Flamanville EPR Timetable and Costs Revised," World Nuclear News, September 3, 2015, https://world-nuclear-news.org/Articles/Flamanville-EPR-timetable-and-costs-revised.

3. Clark, Kevin, "A Nod to the People Who Helped Build Plant Vogtle,"

Power Engineering, May 31, 2024, https://www.power-eng.com/news/a-nod-to-the-people-who-helped-build-plant-vogtle.

4. Dalton, David, "NucNet Explainer: Finland's Olkiluoto-3 Begins Commercial Operation," NucNet, the Independent Nuclear News Agency, May 2, 2023, https://www.nucnet.org/infographics/nucnet-explainer-finland-s-olkiluoto-3-begins-commercial-operation-5-2-2023.

5. "First New U.S. Nuclear Reactor Since 2016 Is Now in Operation," EIA, August 1, 2023, https://www.eia.gov/todayinenergy/detail.php?id=57280.

6. "Energy Policy Act of 2005," Pub. L. No. 109-58, 119 Stat. 594 (2005), https://www1.eere.energy.gov/femp/pdfs/epact_2005.pdf.

7. "Combined License Applications for New Reactors," NRC, July 3, 2023, https://www.nrc.gov/reactors/new-reactors/large-lwr/col.html.

8. "Combined License Holders for New Reactors," NRC, July 21, 2023, https://www.nrc.gov/reactors/new-reactors/large-lwr/col-holder.html.

9. Amy, Jeff, "Timeline: How Georgia and South Carolina Nuclear Reactors Ran So Far off Course," *AP News*, May 25, 2023, https://apnews.com/article/nuclear-power-georgia-vogtle-reactors-8fbf41a3e04c656002a6ee8203988fad.

10. "Heavy Manufacturing of Power Plants," World Nuclear Association, March 4, 2021, https://world-nuclear.org/information-library/nuclear-fuel-cycle/nuclear-power-reactors/heavy-manufacturing-of-power-plants.

11. Nugent, Ciara, "Why Greta Thunberg and Other Climate Activists Are Protesting Wind Farms in Norway," *Time*, February 28, 2023, https://time.com/6259144/greta-thunberg-norway-protests-climate-activists/.

12. "Offshore Wind Industry Hits Rough Waters Amid Rising Costs," *Business Insider*, October 3, 2023, https://markets.businessinsider.com/news/stocks/offshore-wind-industry-hits-rough-waters-amid-rising-costs-1032675804.

13. "Repowering Wind Turbines Adds Generating Capacity at Existing Sites," EIA, November 6, 2017, https://www.eia.gov/todayinenergy/detail.php?id=33632.

14. Bronner, Stephen J., "The Duck Curve: What Is It and Is It a Problem?" *CNET*, n.d., https://www.cnet.com/home/energy-and-utilities/the-duck-curve-what-is-it-and-is-it-a-problem/.

15. "Electric Power Monthly," EIA.
16. Seel, Joachim, Dev Millstein, Andrew Mills, Mark Bolinger, and Ryan Wiser, "Plentiful Electricity Turns Wholesale Prices Negative," *Advances in Applied Energy* 4 (2021), https://doi.org/10.1016/j.adapen.2021.100073.
17. "Electric Power Monthly," EIA.
18. Brock, Anne, "How Do Clouds, Rain & Snow Affect Solar Panel Output?" *Solar Alliance* (blog), January 27, 2022, https://www.solaralliance.com/how-do-clouds-affect-solar-panels/.
19. Gill, Liz, Aleecia Gutierrez, and Terra Weeks, "2021 SB 100 Joint Agency Report, Achieving 100 Percent Clean Electricity in California: An Initial Assessment," California Energy Commission, March 15, 2021, https://www.energy.ca.gov/publications/2021/2021-sb-100-joint-agency-report-achieving-100-percent-clean-electricity.
20. Shea, Daniel, "Nuclear Power and the Clean Energy Transition," National Conference of State Legislatures (NCSL), April 6, 2023, https://www.ncsl.org/energy/nuclear-power-and-the-clean-energy-transition.
21. "Corporate Average Fuel Economy," National Highway Traffic Safety Administration (NHTSA), accessed April 28, 2024, https://www.nhtsa.gov/laws-regulations/corporate-average-fuel-economy.
22. "Federal Financial Interventions and Subsidies in Energy in Fiscal Years 2016–2022," EIA, August 11, 2023, https://www.eia.gov/analysis/requests/subsidy/.
23. Bastiat, Claude Frédéric, *The Bastiat Collection*, 2nd ed., (Auburn, AL: Ludwig von Mises Institute, 2007), https://cdn.mises.org/The%20Bastiat%20Collection_4.pdf.

CHAPTER 6: AN INDUSTRY GASLIGHTED

1. Lovering, Jessica R., Arthur Yip, and Ted Nordhaus, "Historical Construction Costs of Global Nuclear Power Reactors," *Energy Policy* 91 (April): 371–82, https://doi.org/10.1016/j.enpol.2016.01.011.
2. Lovering et al., "Historical Construction Costs of Global Nuclear Power Reactors."
3. Nuclear Energy Agency (NEA), *Legal Challenges Related to Nuclear Safety* (Paris: OECD Publishing, 2024), https://www.oecd-nea.org/

jcms/pl_91598/legal-challenges-related-to-nuclear-safety.

4. Boyle, Elizabeth H., "Political Frames and Legal Activity: The Case of Nuclear Power in Four Countries," *Law & Society Review* 32, no. 3 (1998), pp. 149 and 151.

5. Ruyter, Elena, "Residents Protest, Gain Closure of Vermont Yankee Nuclear Power Plant, USA, 2005–2013," Global Nonviolent Action Database, Swarthmore College, September 25, 2011, https://nvdatabase.swarthmore.edu/content/residents-protest-gain-closure-vermont-yankee-nuclear-power-plant-usa-2005-2013.

6. "Nuclear Power in the USA," World Nuclear Association.

7. Hewett, Frederick, "I've Been Against Nuclear Power for Decades. Until Now," WBUR radio, June 7, 2023. https://www.wbur.org/cognoscenti/2023/06/27/nuclear-power-plants-climate-change-chernobyl-frederick-hewett.

8. Young, Ward, and Mark Evanoff, "The Diablo Canyon Blockade 1981," *FoundSF* (blog), n.d., https://www.foundsf.org/index.php?title=The_Diablo_Canyon_Blockade_1981.

9. Calvert Cliffs' Coordinating Committee, Inc., et al., Petitioners, v. United States Atomic Energy Commission and United States of America, Respondents, 449 F.2d 1109 (D.C. Cir. 1971), Justia Law, April 23, 1971. https://law.justia.com/cases/federal/appellate-courts/F2/449/1109/240994/.

10. Loyola, Mario, "Want to Bring Microchip Fabs Back to the U.S.? Exempt Them from Environmental Review," Heritage Foundation, April 11, 2024, https://www.heritage.org/technology/commentary/want-bring-microchip-fabs-back-the-us-exempt-them-environmental-review.

11. Goldberg, Marshall, "Federal Energy Subsidies: Not All Technologies Are Created Equal," Renewal Energy Policy Project, July 2000, https://www.earthtrack.net/sites/default/files/repp-subsidies.pdf.

12. "NRC Regulations Title 10, Code of Federal Regulations," NRC, n.d., https://www.nrc.gov/reading-rm/doc-collections/cfr/index.html.

13. Delmas, Magali, and Bruce Heiman, "Government Credible Commitment to the French and American Nuclear Power Industries," *Journal of Policy Analysis and Management* 20, no. 3 (Summer 2001), p. 447, http://www.jstor.org/stable/3326131.

14. "Nuclear Powerplant Licensing: Need for Additional Improvements," Report to the Congress of the United States, April 27, 1978, https://

15. www.gao.gov/assets/emd-78-29.pdf.
15. "Summary of the Clean Water Act," EPA, June 22, 2023, https://www.epa.gov/laws-regulations/summary-clean-water-act.
16. Shea, Daniel, "State Options to Keep Nuclear in the Energy Mix," NCSL, May 30, 2017, https://www.ncsl.org/energy/state-options-to-keep-nuclear-in-the-energy-mix.
17. "States Restrictions on New Nuclear Power Facility Construction," NCSL, August 17, 2021, https://www.ncsl.org/environment-and-natural-resources/states-restrictions-on-new-nuclear-power-facility-construction.
18. "The California Nuclear Waste Act [Ballot]," Legislative Analyst's Office, California Legislature, February 26, 2015, https://lao.ca.gov/BallotAnalysis/Initiative/2015-001.
19. Spencer, Jack, "Competitive Nuclear Energy Investment: Avoiding Past Policy Mistakes," Heritage Foundation, November 15, 2007, https://www.heritage.org/environment/report/competitive-nuclear-energy-investment-avoiding-past-policy-mistakes.
20. Potter, Brian, "Why Does Nuclear Power Plant Construction Cost So Much?" Institute for Progress, May 1, 2023, https://ifp.org/nuclear-power-plant-construction-costs/.
21. "Analysis of Nuclear Power Plant Construction Costs," Office of Scientific and Technical Information, DOE, January 1, 1986, https://www.osti.gov/servlets/purl/6071600.
22. Cohen, Bernard L., *The Nuclear Energy Option* (New York: Plenum Press, 1990), chap. 9, www.phyast.pitt.edu/~blc/book/chapter9.html.
23. Potter, "Why Does Nuclear Power Plant Construction Cost So Much?"
24. Cohen, *The Nuclear Energy Option*.
25. Cohen, *The Nuclear Energy Option*.
26. "Fact Sheet on the Proposed Nuclear Non-Proliferation Policy Act of 1977," American Presidency Project, UC Santa Barbara, April 27, 1977, https://www.presidency.ucsb.edu/documents/fact-sheet-the-proposed-nuclear-non-proliferation-policy-act-1977.
27. Bourg, Stéphane, and Christophe Poinssot, "Could Spent Nuclear Fuel Be Considered as a Non-Conventional Mine of Critical Raw Materials?" *Progress in Nuclear Energy* 94 (January 2017): 222–28. https://doi.org/10.1016/j.pnucene.2016.08.004.

28. "All about Used Fuel Processing and Recycling," Orano Group, n.d., https://www.orano.group/en/unpacking-nuclear/all-about-used-fuel-processing-and-recycling.

29. Nuclear Energy Institute, "Plutonium and Uranium Reprocessing," Acamedia, January 2003, www.acamedia.info/politics/nonproliferation/references/nei_2003.htm.

30. Delmas and Heiman, "Government Credible Commitment to the French and American Nuclear Power Industries."

31. Hearth, Douglas, Ronald W. Melicher, and Darryl E.J. Gurley. "Nuclear Power Plant Cancellations: Sunk Costs and Utility Stock Returns," *Quarterly Journal of Business and Economics* 29, no. 1 (1990), https://link.gale.com/apps/doc/A8234077/AONE?u=anon~b2aad448&sid=googleScholar&xid=7ad67728.

32. Carlone, Ralph, "Statement of Ralph v. Carlone, Associate Director Energy and Minerals Division," GAO, July 20, 1978, https://www.gao.gov/assets/095394.pdf.

33. "Energy Policy Act of 2005: Summary and Analysis of Enacted Provisions," March 8, 2006, https://www.everycrsreport.com/reports/RL33302.html.

34. Donalds, Byron, "Donalds Introduces 16th, 17th, 18th, 19th And 20th Bills of 2023–24 Nuclear Energy Package," Office of Congressman Byron Donalds press release, July 18, 2023, https://donalds.house.gov/news/email/show.aspx?ID=7MZIWNT4N4RXW.

35. "Senate Passes Bipartisan Nuclear Energy Bill from Capito, Carper, Whitehouse," US Senate Committee on Environment and Public Works, July 27, 2023, https://www.epw.senate.gov/public/index.cfm/2023/7/senate-passes-bipartisan-nuclear-energy-bill-from-capito-carper-whitehouse.

36. "Advanced Ultra-Supercritical Technology," GE Vernova, n.d., https://www.gevernova.com/steam-power/coal-power-plant/usc-ausc.

CHAPTER 7: GOVERNMENT CAN'T SUBSIDIZE AN INDUSTRY INTO SUCCESS

1. Collins, Hilary, "Nuclear's Renaissance Is Powering Infrastructure Investment," *Middle Market Growth*, March 26, 2024, https://middlemarketgrowth.org/infrastructure-nuclear-investment/.

2. "U.S. State Electricity Portfolio Standards," Center for Climate and

Energy Solutions, January 31, 2020, https://www.c2es.org/document/renewable-and-alternate-energy-portfolio-standards/.

3. "Renewable Energy R&D Funding History: A Comparison with Funding for Nuclear Energy, Fossil Energy, Energy Efficiency, and Electric Systems R&D," Congressional Research Service, June 18, 2018, https://crsreports.congress.gov/product/pdf/RS/RS22858/17.

4. "Energy Tax Act of 1978 – Policies." n.d. IEA. https://www.iea.org/policies/4248-energy-tax-act-of-1978.

5. "U.S. Energy Facts Explained: Consumption and Production," EIA, July 15, 2024, https://www.eia.gov/energyexplained/us-energy-facts/.

6. "Federal Financial Interventions and Subsidies in Energy in Fiscal Years 2016–2022," EIA.

7. Loyola, Mario, "High Electricity Prices Have Europe Facing Deindustrialization; Don't Let It Happen Here," Heritage Foundation, February 12, 2024, https://www.heritage.org/energy/commentary/high-electricity-prices-have-europe-facing-deindustrialization-dont-let-it-happen.

8. "Spencer Abraham Calls for Doubling of US Nuclear Capacity," Nuclear Engineering International, March 2, 2005, https://www.neimagazine.com/news/newsspencer-abraham-calls-for-doubling-of-us-nuclear-capacity.

9. "DOE Loan Programs Office: 2023 Updates, Overview and Key Insights," Holland & Knight, February 1, 2023, https://www.hklaw.com/en/insights/publications/2023/02/doe-loan-programs-office-2023-updates-overview-and-key-insights.

10. "Henry Hub Natural Gas Spot Price," EIA, n.d., https://www.eia.gov/dnav/ng/hist/rngwhhdm.htm.

11. "Design Certification Applications for New Reactors," NRC, May 22, 2023, https://www.nrc.gov/reactors/new-reactors/large-lwr/design-cert.html.

12. "Early Site Permit Applications for New Reactors," NRC, September 21, 2022, https://www.nrc.gov/reactors/new-reactors/large-lwr/esp.html.

13. "Combined License Applications for New Reactors," NRC.

14. "Backgrounder on Nuclear Power Plant Licensing Process," NRC, June 7, 2022, https://www.nrc.gov/reading-rm/doc-collections/fact-sheets/licensing-process-fs.html.

15. Go, Alison, "The New Hot Job: Nuclear Engineering," *U.S. News & World Report*, August 14, 2008, https://www.usnews.com/education/articles/2008/08/14/the-new-hot-job-nuclear-engineering.

16. "UUSA, the National Enrichment Facility: Key Facts," Urenco, accessed June 14, 2024, https://www.urenco.com/global-operations/uusa.

17. "Nuclear Engineering Enrollments and Degrees Survey, 2006 Data," Oak Ridge Institute for Science and Education, 2007, https://www.osti.gov/servlets/purl/937436.

18. "VA Nuclear Manufacturing Plant Delayed," Manufacturing.net, August 19, 2010, https://www.manufacturing.net/home/news/13133637/va-nuclear-manufacturing-plant-delayed.

19. Modany, Angela, "Pong, Atari, and the Origins of the Home Video Game," National Museum of American History, April 17, 2012, https://americanhistory.si.edu/blog/2012/04/pong-atari-and-the-origins-of-the-home-video-game.html.

20. "50 Years of Video Game Industry Revenues, by Platform," Visual Capitalist, December 31, 2023, https://www.visualcapitalist.com/video-game-industry-revenues-by-platform/.

21. Morton, Wes, "A $406bn Gaming Industry and the Melee for Advertiser's next Frontier," *The Drum*, November 17, 2023, https://www.thedrum.com/opinion/2023/11/17/406bn-gaming-industry-and-the-melee-advertisers-next-frontier.

22. Dayaratna, Kevin, and Nicolas Loris, "Assessing the Costs and Benefits of the Green New Deal's Energy Policies," Heritage Foundation, July 24, 2019, https://www.heritage.org/sites/default/files/2019-07/BG3427.pdf. The report finds the policy would lead to an annual loss of $8,000 for families and 1.2 million jobs per year through 2040.

23. Tubb, Katie, Nicolas Loris, and Rachel Zissimos, "Taking the Long View: How to Empower the Coal and Nuclear Industries to Compete and Innovate," Heritage Foundation, September 15, 2018, https://www.heritage.org/energy-economics/report/taking-the-long-view-how-empower-the-coal-and-nuclear-industries-compete.

CHAPTER 8: THE PROMISE, CHALLENGE, AND OPPORTUNITY OF ADVANCED REACTORS

1. "Ontario Power Generation and Synthos Green Energy Invest in Development of GE Hitachi Small Modular Reactor Technology,"

Tennessee Valley Authority, March 23, 2023, https://www.tva.com/newsroom/press-releases/tennessee-valley-authority-ontario-power-generation-and-synthos-green-energy-invest-in-development-of-ge-hitachi-small-modular-reactor-technology. See also "NEI Advanced Nuclear Demand Survey," NEI, 2022, https://www.nei.org/CorporateSite/media/filefolder/news/Advanced-Nuclear-Demand-Survey.pdf.

2. Wald, Matthew L., "Alaska Town Seeks Reactor to Cut Costs of Electricity." *New York Times*, February 3, 2005, https://www.nytimes.com/2005/02/03/business/alaska-town-seeks-reactor-to-cut-costs-of-electricity.html.

3. Cameron, Susan, "Gilbert and West Virginia Counterpart Announce Push for Small Nuclear Reactors in Rural Areas," *Cardinal News*, December 1, 2022, https://cardinalnews.org/2022/10/13/gilbert-and-west-virginia-counterpart-announce-push-for-small-nuclear-reactors-in-rural-areas/.

4. IEA, IRENA, UNSD, World Bank, WHO, *Tracking SDG 7: The Energy Progress Report 2023*, (Washington: World Bank, 2023), https://iea.blob.core.windows.net/assets/9b89065a-ccb4-404c-a53e-084982768baf/SDG7-Report2023-FullReport.pdf.

5. IEA et al., *Tracking SDG 7*.

6. Deign, Jason, "These Countries Are Investing More in Small Modular Reactors in the New Decade," World Economic Forum, January 13, 2021, https://www.weforum.org/agenda/2021/01/buoyant-global-outlook-for-small-modular-reactors-2021/.

7. "Global Nuclear SMR Project Pipeline Expands to 22 GW, Increasing More than 65% since 2021," Wood Mackenzie press release, March 7, 2024, https://www.woodmac.com/press-releases/2024-press-releases/global-nuclear-smr-project-pipeline-expands-to-22-gw-increasing-more-than-65-since-2021/.

8. "Advanced Nuclear Solutions," Tennessee Valley Authority, n.d., https://www.tva.com/energy/technology-innovation/advanced-nuclear-solutions. See also "Advanced Nuclear Reactor Project in Seadrift, Texas," Dow and X-energy, n.d., https://x-energy.com/seadrift.

9. Allen, Wendy, "Nuclear Reactors for Generating Electricity: U.S. Development from 1946 to 1963," Santa Monica, CA: RAND Corporation, 1977, https://www.rand.org/pubs/reports/R2116.html.

10. "Power Reactors, Oak Ridge, Tenn," Technical Information Service, US Atomic Energy Commission, 1958.

11. US commercial nuclear power primarily uses light-water reactors: boiling-water reactors (BWRs) and pressurized-water reactors (PWRs). BWRs generate steam directly from water heated by fission, while PWRs use pressurized water to heat a separate water source for steam. High-temperature gas-cooled reactors (HTGRs) differ by using helium gas heated to very high temperatures in a graphite-moderated core to power gas turbines or provide process heat.

12. Smith, Rebecca, "Small Reactors Generate Big Hopes," *Wall Street Journal*, February 18, 2010, https://www.wsj.com/articles/SB10001424052748703444804575071402124482176.

13. "NRC Certifies First U.S. Small Modular Reactor Design," DOE press release, January 20, 2023, https://www.energy.gov/ne/articles/nrc-certifies-first-us-small-modular-reactor-design.

14. "Licensing Activities: Pre-Application Activities," NRC, October 9, 2023, https://www.nrc.gov/reactors/new-reactors/advanced/who-were-working-with/licensing-activities/pre-application-activities/genatom.html.

15. "Small Nuclear Power Reactors," World Nuclear Association, February 16, 2024, https://world-nuclear.org/information-library/Nuclear-Fuel-Cycle/Nuclear-Power-Reactors/small-nuclear-power-reactors.

16. Rickover, Hyman, "Admiral Rickover's 'Paper Reactor' Memo," *What Is Nuclear?*, June 5, 1953, https://whatisnuclear.com/rickover.html.

17. It should be noted that some argue Rickover's decision to force the navy and the nation into light-water technology was not as successful as it is often portrayed. More technological competition may have yielded a more affordable reactor technology that would allow nuclear to power naval vessels across the fleet rather than only on submarines and aircraft carriers.

18. "Design Certification: Nuscale US600," NRC, March 14, 2024, https://www.nrc.gov/reactors/new-reactors/smr/licensing-activities/nuscale.html; "SMR Pre-Application Activities," NRC, March 14, 2024, https://www.nrc.gov/reactors/new-reactors/smr/licensing-activities/pre-application-activities.html; "Aurora–Oklo Application," NRC, September 18, 2022, https://www.nrc.gov/reactors/new-reactors/large-lwr/col/aurora-oklo.html; "SMR Pre-Application Activities," NRC, 2022, https://www.nrc.gov/reactors/new-reactors/smr/licensing-activities/pre-application-activities.html.

19. S.512: Nuclear Energy Innovation and Modernization Act, Pub. L. 115-439 (2017), https://www.congress.gov/bill/115th-congress/senate-bill/512.

20. Wagner, John, NAC International, Ben Cipiti, and R. N. Blomquist, "Developing a New Regulatory Framework for Advanced Reactors: Update on Part 53," *NuclearNewswire*, May 3, 2024, https://www.ans.org/news/article-6003/developing-a-new-regulatory-framework-for-advanced-reactor.

21. "Nuclear Reactor Construction: Starts Drop Again in the World," World Nuclear Industry Status Report, March 18, 2024, https://www.worldnuclearreport.org/Nuclear-Reactor-Construction-Starts-Drop-Again-in-the-World.html.

22. "Preparatory Work Stepped Up for Russia's First Land-Based SMR," *World Nuclear News*, February 2, 2024, https://www.world-nuclear-news.org/Articles/Preparatory-work-stepped-up-for-Russia-s-first-lan#.

23. "Russia Set to Build SMR Nuclear Power Plant in Uzbekistan," *World Nuclear News*, May 28, 2024, https://www.world-nuclear-news.org/Articles/Russia-set-to-build-SMR-nuclear-power-plant-in-Uzb.

24. "Plans for New Reactors Worldwide," World Nuclear Association, April 30, 2024, https://world-nuclear.org/information-library/current-and-future-generation/plans-for-new-reactors-worldwide.

25. "China Continues Rapid Growth of Nuclear Power Capacity," EIA, May 6, 2024, https://www.eia.gov/todayinenergy/detail.php?id=61927.

26. Li, Aitong, Yahan Liu, and Zongyao Yu, "China's Nuclear Exports: Understanding the Dynamics between Domestic Governance Reforms and International Market Competition," *Energy Research & Social Science* 103 (2023), https://doi.org/10.1016/j.erss.2023.103230.

CHAPTER 9: FUELING THE TWENTY-FIRST CENTURY

1. "Where Our Uranium Comes From," EIA, August 23, 2023, https://www.eia.gov/energyexplained/nuclear/where-our-uranium-comes-from.php.

2. "U.S. Uranium Production Up in 2022 After Reaching Record Lows in 2021," EIA, August 17, 2023, https://www.eia.gov/todayinenergy/detail.php?id=60160.

3. "Uranium Enrichment," World Nuclear Association, October 11, 2022, https://world-nuclear.org/information-library/nuclear-fuel-cycle/conversion-enrichment-and-fabrication/uranium-enrichment.

4. "Nuclear Share of Electricity Generation in 2023," Power Reactor Information System (PRIS), International Atomic Energy Agency (IAEA), October 14, 2024, https://pris.iaea.org/PRIS/WorldStatistics/NuclearShareofElectricityGeneration.aspx.

5. "U.S. Uranium Concentrate Production in 2021 Remained Near All-Time Lows," EIA, July 26, 2022, https://www.eia.gov/todayinenergy/detail.php?id=53179.

6. "U.S. Uranium Production Fell to an All-Time Annual Low in 2019," EIA, July 17, 2020, https://www.eia.gov/todayinenergy/detail.php?id=44416.

7. "US Nuclear Fuel Cycle," World Nuclear Association, November 6, 2023, https://world-nuclear.org/information-library/country-profiles/countries-t-z/usa-nuclear-fuel-cycle.

8. "World Uranium Mining Production," World Nuclear Association, April 30, 2024, https://world-nuclear.org/information-library/nuclear-fuel-cycle/mining-of-uranium/world-uranium-mining-production.

9. "Domestic Uranium Production Report: Annual," EIA, 2024, https://www.eia.gov/uranium/production/annual/udrilling.php.

10. "Uranium Markets," World Nuclear Association, May 7, 2024, https://world-nuclear.org/information-library/nuclear-fuel-cycle/uranium-resources/uranium-markets.

11. Jamasmie, Cecilia, "Uranium Price Jumps to 15-Year High as Top Miner Flags Shortfall," Mining.com, January 12, 2024, https://www.mining.com/uranium-jumps-to-15-year-high-as-top-miner-flags-shortfall/.

12. "Uranium," Trading Economics, 2024, https://tradingeconomics.com/commodity/uranium.

13. "The Coles Hill Uranium Deposit, Virginia, USA: Geology, Geochemistry, Geochronology, and Genetic Model," US Geological Survey, March 1, 2022, https://www.usgs.gov/publications/coles-hill-uranium-deposit-virginia-usa-geology-geochemistry-geochronology-and-genetic.

14. "Establishment of the Baaj Nwaavjo I'tah Kukveni Ancestral Footprints of the Grand Canyon National Monument," *Federal Register*, National Archives and Records Administration,

August 15, 2023, https://www.federalregister.gov/documents/2023/08/15/2023-17628/establishment-of-the-baaj-nwaavjo-itah-kukveni-ancestral-footprints-of-the-grand-canyon-national.

15. Tubb, Katie, "There's No Need to Subsidize American Uranium," *Daily Signal*, June 18, 2019, https://www.dailysignal.com/2019/06/18/theres-no-need-to-subsidize-american-uranium/.

16. Heritage Foundation, "Powering America: Uranium Mining and Milling," YouTube, 2012, https://www.youtube.com/watch?v=oT2LHGG-9Ko.

17. "Uranium Recovery," NRC, March 31, 2023, https://www.nrc.gov/materials/uranium-recovery.html.

18. Holzman, Jael, and Hannah Northey, "Biden Wants Minerals, but Mine Permitting Lags," *E&E News by Politico*, August 9, 2022, https://www.eenews.net/articles/biden-wants-minerals-but-mine-permitting-lags/.

19. "Testimony of Rich Nolan, National Mining Association: Hearing to Examine Opportunities for Congress to Reform the Permitting Process for Energy and Mineral Projects," US Senate Energy & Natural Resources Committee, May 11, 2023, https://nma.org/wp-content/uploads/2023/05/National-Mining-Association-5-11-23-Nolan-Testimony.pdf.

20. "Domestic Uranium Production Report—Quarterly," EIA, February 9, 2023, https://www.eia.gov/uranium/production/quarterly/.

21. "Nuclear Explained: Where Our Uranium Comes From," EIA, July 7, 2022, https://www.eia.gov/energyexplained/nuclear/where-our-uranium-comes-from.php.

22. Patel, Sonal, "Honeywell to Reopen Sole U.S. Uranium Conversion Plant," *POWER* magazine, February 9, 2021, https://www.powermag.com/honeywell-to-reopen-sole-u-s-uranium-conversion-plant/. The US DOE has announced that it will purchase $14M in conversion services. Jennetta, Andrea, "US DOE Awards $14 Million to US Uranium Converter for Reserve Program," SP Global, January 5, 2023, https://www.spglobal.com/commodityinsights/en/market-insights/latest-news/electric-power/010523-us-doe-awards-14-million-to-us-uranium-converter-for-reserve-program.

23. "Nuclear Fuel: Metropolis Sold Out through 2028, Weighing Expansion," Energy Intelligence, November 17, 2023. https://www.

energyintel.com/0000018b-d951-dbb5-a5ef-dd7328030000.

24. "UUSA: Key Facts," Urenco.
25. "US Nuclear Fuel Cycle," World Nuclear Association.
26. "Uranium Marketing Annual Report," EIA, February 9, 2023, https://www.eia.gov/uranium/marketing/table16.php.
27. "Urenco's First Capacity Expansion to Be at Its US Site," Urenco, July 6, 2023, https://www.urenco.com/news/global/2023/urencos-first-capacity-expansion-to-be-at-its-us-site.
28. Hinshaw, Drew, and Joe Parkinson, "Russia Gives Ukraine Nuclear Plant Workers Ultimatum to Pick a Side," *Wall Street Journal*, October 17, 2022, https://www.wsj.com/articles/russia-gives-ukraine-nuclear-plant-workers-ultimatum-to-pick-a-side-11666025113.
29. "Megatons to Megawatts," Centrus Energy Corp, June 28, 2016, https://www.centrusenergy.com/who-we-are/history/megatons-to-megawatts/.
30. "ICYMI: Barrasso Op-Ed: We Must End American Dependence on Russian Uranium," US Senate Committee on Energy and Natural Resources, March 21, 2022, https://www.energy.senate.gov/2022/3/icymi-barrasso-op-ed-we-must-end-american-dependence-on-russian-uranium.
31. Prohibiting Russian Uranium Imports Act, Pub. L. No. 118-62 (2023), https://www.congress.gov/bill/118th-congress/house-bill/1042/text.
32. "Urenco's First Capacity Expansion to Be at Its US Site," Urenco.
33. "Testimony of Rich Nolan, National Mining Association," US Senate Energy & Natural Resources Committee.
34. "FY2024 Spending Bill Fuels Historic Push for U.S. Advanced Reactors," Office of Nuclear Energy, DOE, March 14, 2024, https://www.energy.gov/ne/articles/fy2024-spending-bill-fuels-historic-push-us-advanced-reactors.
35. "US Nuclear Fuel Cycle," World Nuclear Association.
36. "Uranium Marketing Annual Report," EIA, June 6, 2024, https://www.eia.gov/uranium/marketing/.
37. Price, Rowen, Ryan Norman, and Alan Ahn, "Western Reliance on Russian Fuel: A Dangerous Game," *Third Way*, September 20, 2023, https://www.thirdway.org/memo/western-reliance-on-russian-fuel-a-dangerous-game.
38. Vlasov, Artem, "Thorium's Long-Term Potential in Nuclear Energy:

New IAEA Analysis," International Atomic Energy Agency (IAEA), March 13, 2023, https://www.iaea.org/newscenter/news/thoriums-long-term-potential-in-nuclear-energy-new-iaea-analysis.

39. Hanson, Maris, "Uranium 233: The Nuclear Superfuel No One Is Using," *Seattle Journal of Technology, Environmental & Innovation Law* 12, no. 1 (2022): 1–19, https://digitalcommons.law.seattleu.edu/cgi/viewcontent.cgi?article=1030&context=sjteil.

40. "Radioisotopes in Medicine," World Nuclear Association, April 30, 2024, https://world-nuclear.org/information-library/non-power-nuclear-applications/radioisotopes-research/radioisotopes-in-medicine.

41. "Shippingport Light Water Breeder Reactor Remarks at a Ceremony Marking the Pennsylvania Facility's Increase to Full Power Production," American Presidency Project, UC Santa Barbara, December 2, 1977, https://www.presidency.ucsb.edu/documents/shippingport-light-water-breeder-reactor-remarks-ceremony-marking-the-pennsylvania.

42. "Thorium," World Nuclear Association, May 2, 2024, https://world-nuclear.org/information-library/current-and-future-generation/thorium.

43. Firsing, Scott, "How China Has Pulled Ahead in the Global Nuclear Race," *International Policy Digest*, December 4, 2023, https://intpolicydigest.org/how-china-has-pulled-ahead-in-the-global-nuclear-race/.

44. Huotari, John, "DOE Disposing of Uranium-233 Waste Stored at ORNL," *Oak Ridge Today*, August 27, 2017, https://oakridgetoday.com/2017/08/27/doe-program-disposing-uranium-233-waste-stored-ornl/.

45. "123 Agreements: Fact Sheet," US Department of State, December 6, 2022, https://www.state.gov/fact-sheets-bureau-of-international-security-and-nonproliferation/123-agreements/.

CHAPTER 10: THE ATOMIC OPPORTUNITY OF NUCLEAR WASTE

1. Schneider, Mycle, and Yves Marignac, "Spent Nuclear Fuel Reprocessing in France," International Panel on Fissile Materials, April 2008, https://fissilematerials.org/library/rr04.pdf.

2. "All about Used Fuel Processing and Recycling," Orano Group.
3. Krikorian, Shant, "France's Efficiency in the Nuclear Fuel Cycle: What Can 'Oui' Learn?" IAEA, September 4, 2019, https://www.iaea.org/newscenter/news/frances-efficiency-in-the-nuclear-fuel-cycle-what-can-oui-learn.
4. Nuclear Waste Policy Act of 1982, Pub. L. No. 97-425, 96 Stat. 2201 (1983), https://www.energy.gov/articles/nuclear-waste-policy-act.
5. *Agency Financial Report: Fiscal Year 2023*, DOE, 2023. https://www.energy.gov/sites/default/files/2023-11/fy-2023-doe-agency-financial-report_0.pdf.
6. *Commercial Spent Nuclear Fuel: Congressional Action Needed to Break Impasse and Develop a Permanent Disposal Solution*, GAO (GAO-21-603), September 2021, https://www.gao.gov/assets/gao-21-603.pdf.
7. "Nuclear Waste Disposal," GAO, n.d., https://www.gao.gov/nuclear-waste-disposal. See also "Nuclear Waste Storage Sites in the United States," Congressional Research Service, April 13, 2020, https://sgp.fas.org/crs/nuke/IF11201.pdf.
8. *Report to the President and Congress by the Secretary of Energy on the Need for a Second Repository*, DOE (DOE/RW05-95), December 2008, https://www.energy.gov/articles/report-president-and-congress-secretary-energy-need-second-repository.
9. "Room at the Mountain," Economic Power Research Institute, 2007, http://mydocs.epri.com/docs/CorporateDocuments/SectorPages/Portfolio/Nuclear/YuccaMtn.html.
10. Brady, Jeff, "As Nuclear Waste Piles Up, Private Companies Pitch New Ways to Store It," *NPR*, April 30, 2019, https://www.npr.org/2019/04/30/716837443/as-nuclear-waste-piles-up-private-companies-pitch-new-ways-to-store-it.
11. *Blue Ribbon Commission on America's Nuclear Future: Report to the Secretary of Energy*, Blue Ribbon Commission on America's Nuclear Future, January 2012, https://www.energy.gov/sites/prod/files/2013/04/f0/brc_finalreport_jan2012.pdf.
12. Holdsworth, Alistair F., Harry Eccles, Clint A. Sharrad, and Kathryn George, "Spent Nuclear Fuel—Waste or Resource? The Potential of Strategic Materials Recovery During Recycle for Sustainability and Advanced Waste Management," *Waste* 1, no. 1 (January 15, 2023), https://doi.org/10.3390/waste1010016.
13. Willson, Miranda, "Some in Nye County Say Yucca Mountain Could

Be a Blessing. Others Aren't Convinced," *Las Vegas Sun*, May 14, 2019, https://lasvegassun.com/news/2019/may/14/some-in-nye-county-willing-listen-yucca-mountain/.

14. Generic Environmental Impact Statement for License Renewal of Nuclear Plants: Regarding Indian Point Energy Center, Units 2 and 3," NRC, January 2012, https://www.nrc.gov/reading-rm/doc-collections/nuregs/staff/sr1437/supplement38/index.html.

CHAPTER 11: A POLICY REVOLUTION

1. "Inflation-Adjusted U.S. Energy Spending Increased by 25% in 2021," EIA, August 3, 2023, https://www.eia.gov/todayinenergy/detail.php?id=57320. See also *Strategies for Affordable and Fair Clean Energy Transitions*, EIA, May 2024, https://www.iea.org/reports/strategies-for-affordable-and-fair-clean-energy-transitions.

2. "Radiation Regulations and Laws," EPA, December 7, 2023, https://www.epa.gov/radiation/radiation-regulations-and-laws.

3. "EPA Is Taking Steps to Update Its Federal Radiation Guidance," Oversight.gov, January 6, 2022, https://www.oversight.gov/report/EPA/Hotline-EPA-Taking-Steps-Update-Its-Federal-Radiation-Guidance.

4. Baran, Jeffery M., "Policy Issue Notation Vote," NRC, May 14, 2021, https://www.nrc.gov/docs/ML2123/ML21230A138.pdf.

5. Calabrese, Edward J., Jaap C. Hanekamp, and Dima Yazji Shamoun, "The EPA Cancer Risk Assessment Default Model Proposal: Moving Away From the LNT," *Dose Response* 16, 3 (August 9, 2018), https://doi.org/10.1177/1559325818789840.

6. Sykes, Pamela J., "The Benefits of Low Dose Radiation and the Hazards of the Media," *Pathology* 44 (2012), https://doi.org/10.1016/s0031-3025(16)32610-1.

7. "Radiation Sources and Doses," EPA, February 22, 2024, https://www.epa.gov/radiation/radiation-sources-and-doses.

8. Dobrzyński, Ludwik, Krzysztof W. Fornalski, and Ludwig E. Feinendegen, "Cancer Mortality among People Living in Areas with Various Levels of Natural Background Radiation," *Dose Response* 13, 3 (July 2, 2015), https://doi.org/10.1177/155932581559239.

9. "EPA Is Taking Steps to Update Its Federal Radiation Guidance," Oversight.gov.

10. "EPA Is Taking Steps to Update Its Federal Radiation Guidance," Oversight.gov.
11. Kriebel, D., J. Tickner, P. Epstein, J. Lemons, R. Levins, E. L. Loechler, M. Quinn, R. Rudel, T. Schettler, and M. Stoto, "The Precautionary Principle in Environmental Science," *Environmental Health Perspectives* 109, 9 (September 2001), https://www.ncbi.nlm.nih.gov/pmc/articles/PMC1240435/.
12. "ALARA: Definition," NRC, March 9, 2021, https://www.nrc.gov/reading-rm/basic-ref/glossary/alara.html.
13. "Part 20: Standards for Protection Against Radiation," NRC, November 8, 2006, https://www.nrc.gov/reading-rm/doc-collections/cfr/part020/full-text.html.
14. "ALARA: Definition," NRC.
15. "Agreement State Program," NRC, February 8, 2024, https://www.nrc.gov/about-nrc/state-tribal/agreement-states.html.
16. "Special Nuclear Material," NRC, February 6, 2023, https://www.nrc.gov/materials/types/sp-nucmaterials.html#why.
17. "Pre-Application Activities for Advanced Reactors," NRC, July 15, 2024, https://www.nrc.gov/reactors/new-reactors/advanced/who-were-working-with/pre-application-activities.html.
18. "Part 53: Risk Informed, Technology-Inclusive Regulatory Framework for Advanced Reactors, NRC, October 9, 2024, https://www.nrc.gov/reactors/new-reactors/advanced/modernizing/rulemaking/part-53.html.
19. "Capito, Carper, McMorris Rodgers, Pallone Lead Bipartisan, Bicameral Effort Urging NRC to Establish Useable Advanced Nuclear Reactor Licensing Framework," US Senate Committee on Environment & Public Works, July 17, 2023, https://www.epw.senate.gov/public/index.cfm/2023/7/capito-carper-mcmorris-rodgers-pallone-lead-bipartisan-bicameral-effort-urging-nrc-to-establish-useable-advanced-nuclear-reactor-licensing-framework. See also "Groups Call on US Regulator to Finalise Advanced Reactor Rules," World Nuclear News, August 28, 2023, https://world-nuclear-news.org/Articles/Groups-call-on-US-regulator-to-finalise-advanced-r.
20. "X-energy and Ares Acquisition Corporation Mutually Agree to Terminate Business Combination Agreement," X-energy press release, October 31, 2023, https://x-energy.com/media/news-releases/x-energy-ares-mutually-terminate-business-agreement. See also "Utah Associated Municipal Power Systems (UAMPS) and NuScale Power

Agree to Terminate the Carbon Free Power Project (CFPP)," NuScale Power press release, November 8, 2023. https://www.nuscalepower.com/en/news/press-releases/2023/uamps-and-nuscale-power-agree-to-terminate-the-carbon-free-power-project.

21. "Light Water Reactor Sustainability (LWRS) Program," Office of Nuclear Energy, DOE, n.d., https://www.energy.gov/ne/light-water-reactor-sustainability-lwrs-program.

22. "What's the Lifespan for a Nuclear Reactor? Much Longer Than You Might Think," Office of Nuclear Energy, DOE, April 16, 2020, https://www.energy.gov/ne/articles/whats-lifespan-nuclear-reactor-much-longer-you-might-think.

23. "The Future of Nuclear Power: 2023 Baseline Survey," Nuclear Energy Institute (NEI), 2023, https://www.nei.org/CorporateSite/media/filefolder/advantages/The-Future-of-Nuclear-Power-2023-Baseline-Survey.pdf.

24. Pearl, Larry, "Holtec International Formally Launches Process to Restart Palisades Nuclear Power Plant," *Utility Dive*, September 13, 2023, https://www.utilitydive.com/news/palisades-nuclear-holtec-wolverine-hoosier-power-purchase-ppa/693480/.

25. "Constellation to Launch Crane Clean Energy Center, Restoring Jobs and Carbon-Free Power to the Grid," Constellation Energy press released, September 20, 2024, https://www.constellationenergy.com/newsroom/2024/Constellation-to-Launch-Crane-Clean-Energy-Center-Restoring-Jobs-and-Carbon-Free-Power-to-The-Grid.html.

26. "PA Statewide Results on Consumer Attitudes Towards Nuclear Power," Susquehanna Polling and Research press release, August 19, 2024, https://www.abc27.com/wp-content/uploads/sites/55/2024/09/PollMemo-PAStatewide-Omnibus-SRA-August2024.pdf.

27. "AI Is Poised to Drive 160% Increase in Data Center Power Demand," Goldman Sachs, May 14, 2024, https://www.goldmansachs.com/insights/articles/AI-poised-to-drive-160-increase-in-power-demand.

28. Howland, Ethan, "Talen-Amazon Interconnection Agreement Needs Extended FERC Review: PJM Market Monitor," *Utility Dive*, July 11, 2024, https://www.utilitydive.com/news/talen-amazon-interconnection-agreement-ferc-constellation-vistra/721066/.

29. "Nuclear Power: NRC Needs to Take Additional Actions to Prepare to License Advanced Reactors," GAO (GAO 23-105997), July 2023, https://www.gao.gov/assets/d23105997.pdf.

30. Burdick, Stephen J., John C. Wagner, and Jess C. Gehin, "Recommendations to Improve the Nuclear Regulatory Commission Reactor Licensing and Approval Process," Idaho National Laboratory, April 2023, https://inldigitallibrary.inl.gov/sites/sti/sti/Sort_65730.pdf.

31. "The Price-Anderson Act: 2021 Report to Congress," NRC (NUREG/CR-7293), December 2021, https://www.nrc.gov/docs/ML2133/ML21335A064.pdf.

32. "The Price-Anderson Act: 2021 Report to Congress," NRC (NUREG/CR-7293).

33. "The Price-Anderson Act: 2021 Report to Congress," NRC (NUREG/CR-7293).

34. "Insurers Can Help Improve the Image of Nuclear," World Nuclear News, September 16, 2024, https://world-nuclear-news.org/Articles/Insurers-can-help-improve-the-image-of-nuclear.

35. "Uranium Enrichment," World Nuclear Association.

36. "U.S. Uranium Concentrate Production in 2021 Remained Near All-Time Lows," EIA.

37. "U.S. Uranium Production Fell to an All-Time Annual Low in 2019," EIA. See also "US Nuclear Fuel Cycle," World Nuclear Association.

38. Bearak, Max, "The U.S. Is Paying Billions to Russia's Nuclear Agency. Here's Why," *New York Times*, June 14, 2023, https://www.nytimes.com/2023/06/14/climate/enriched-uranium-nuclear-russia-ohio.html.

39. Joselow, Maxine, "U.S. Bans Russian Uranium Imports, Key to Nuclear Fuel Supply," *Washington Post*, May 13, 2024, https://www.washingtonpost.com/business/2024/05/13/russian-uranium-imports-ban/.

40. "Urenco's First Capacity Expansion to Be at Its US Site," Urenco press release, July 6, 2023, https://www.urenco.com/news/global/2023/urencos-first-capacity-expansion-to-be-at-its-us-site.

41. "Testimony of Rich Nolan, National Mining Association," US Senate Energy & Natural Resources Committee.

INDEX

Note: Page numbers followed by "*f*" refer to figures.

Abalone Alliance, 71
Abraham, Spencer, 88
ADVANCE Act, 78
advanced reactors, *see* small modular reactors (SMRs)
Agreement State Program, 151
air quality (US), 3
ALARA (as low as reasonably achievable), 150
Alaska, 104
Amazon Web Services, 155–56
American Centrifuge Project, 91
American nuclear power, a revolution, xiii–xiv
antinuclear litigation (1970s), 69–71
antinuclear movement, 69–71, 130, 134
Atomic Energy Act (AEA) (1954), 151–52
Atomic Energy Advancement Act, 78
Atomic Energy Commission, 102
Australia, 112
auto industry, electrifying, 7

Baaj Nwaavjo I'tah Kukveni, 114–15
Babcock & Wilcox (now BWXT), 104

Barrasso, John, 121
Bastiat, Frédéric, 66–67
battery technology, 62
Belt and Road Initiative, 111
Biden, Joe, 122, 123, 162
    administration, 114–16
    CO_2 reduction goals, 14, 16–17
    role for nuclear power, 24
Bipartisan Infrastructure Bill, 107
Blue Ribbon Commission on America's Nuclear Future (BRC), 136, 141
boiling water reactors (BWRs), 102–03
Brazil, 26
Breitbart, Andrew, xi
Burr, Richard, 88

California, 25, 52, 61
    renewable portfolio standards, 62
Canada, 104, 112, 123
cancers, 46–47
    U-233, 125, 126
carbon dioxide emissions, 4, 24–25, 32
    reductions, 14–17, 95–96
Carter, Jimmy, 76, 77

Chernobyl accident (1986), xi, 12, 13, 41, 46–48, 50
Cherokee plant (South Carolina), 77–78
China, 78, 123, 131, 156
    advanced reactor market, 102, 110–11
    nuclear energy and national security, 28, 29, 30
    nuclear fuel, recycling, 77
    nuclear reactors, 11, 15, 27
    rare earth materials, 26
    thorium research, 126
Chu, Steven, 88
Clamshell Alliance, 71
climate change, 3–4
climate-related deaths, 3
coal plants, 22
coal power, 8
coal reserves, 6, 10
Cohen, Bernard, 75
Cold War, 113, 119, 162
combined licenses (COLs), 56–57
commercial nuclear energy, 18–30
    advancing national and energy security, 26–30
    affordability, 18–24, 20*f*, 21*f*, 24*f*, 82, 85
    environmental considerations, 24–26
    new framework for, 146
    overregulation and decline of industry, 75–78, 76*f*
community colleges, 92
computer gaming, 94
Consolidated National Interveners, 69
cost overruns, 19, 55–62
    overregulation and, 75–78, 76*f*
crude oil reserves, 5–6
Crude Oil Windfall Profit Tax Act (1980), 87
Crystal River 3 nuclear plant (Florida), 70

Davis-Besse nuclear plant (Ohio), 42
Dayaratna, Kevin, 4, 95
decarbonization, 87
deindustrialization, 87
Democrats, xi
Department of Energy (DOE), 5, 38, 105, 109–10, 112, 123, 132, 133, 134, 136, 137, 140, 142, 143
Department of Energy Organization Act (1977), 5, 6
desalination plants, 101
Diablo Canyon nuclear plant (California), 52, 71
District of Columbia, 62
Donalds, Byron, 78
Dresden 1 nuclear plant (Illinois), 102–03
Duane Arnold nuclear plant (Iowa), 70
"duck tail curve," 59–60, 60*f*
Duke Energy, 92

early site permits (ESPs), 89–90
earthquakes, 42–43, 48, 50, 51
Economic Security and Recovery Act (2001), 87
Edison Welding Institute, 91
efficiency mandates, 63–64
electric cars, 32
electricity demand, global, 4
electricity grid, instability in, 85–86, 86*f*
energy crisis (1970s), 5, 86
energy demands, 1
energy industry, and government intervention, 7–11
Energy Information Administration (EIA), 5–6, 19, 65
Energy Multiplier Module (EM2) reactor, 104–05
Energy Policy Act (1992), 78, 87, 120
Energy Policy Act (EPACT) (2005), 17, 56, 57, 79, 87, 88, 89, 90, 105, 107
    failure of, 92–95
energy policy, 4–6, 19, 26, 40, 82–83, 88, 96
    auto industry, electrifying, 7
    nuclear fuel policy reform, market principles to, 161–62
    see also policy revolution
energy shortfalls, 4, 5, 8
energy subsidies, 25, 40, 61, 62, 81–98

ending, 146–48
politicians' reasons to support, 82
Energy Tax Act (1978), 87
Environmental Progress, 13
Environmental Protection Agency (EPA), 3, 34, 148–49
Epstein, Alex, 1
EU Emissions Trading System, 8
European Union (EU)
    energy reserves, 10
    nuclear power, 10
    renewable energy mandates, 87

fabrication, 117
"fake news," *see* myths (nuclear power)
fighter jets, 94
Finland, 13, 99
Flamanville 3 reactor (France), 55
FLEX, 51
Florence-Darlington Technical College's Advanced Welding and Cutting Center, 92
Forbes, Price, 159
Ford, Gerald, 76
Fort Calhoun nuclear plant (Nebraska), 70
Fort Saint Vrain nuclear plant (Colorado), 125
Fossil Future (Epstein), 1
France, 33, 69, 77, 78, 99, 117, 123
    nuclear reactors, 10
    nuclear waste, 130–31
free energy markets, 6–7
Fukushima accident (2011), xi, 12, 39, 41, 42, 48–50, 51, 52

Galena (Alaska), 100
gasoline prices, 8
General Atomics, 104–05
Georgia Power, 57
German Chambers of Commerce and Industry, 8
Germany, 52, 87, 162
    energy policy, 8–9, 10

Global Laser Enrichment facility (North Carolina), 91
Goldman Sachs, 155
Government Accountability Office (GAO), 156, d157
Grand Canyon, 114–16
Grand Canyon National Park, 115
Grays Harbor (Washington), 104

Heritage Foundation, 17
human well-being and energy, 2
Hurricane Katrina, 159
hydrocarbons, 2–3, 9, 19, 27
    air quality, 3
    climate change, 3–4

Idaho Falls enrichment facility (Idaho), 91
Idaho National Laboratory, 157
India, 37
    nuclear reactors, 11, 27
Indian Point 2 nuclear plant (New York), 70
Indian Point 3 nuclear plant (New York), 70
Industrial Revolution, 2
Inflation Reduction Act (IRA), 40, 59, 84, 87, 98, 107
Institute for Energy Research (IER), 6
Institute for Nuclear Power Operations, 46
Institute of Nuclear Power Operations (INPO), 46, 53
International Atomic Energy Agency, 46
International Renewable Energy Agency, 13
investors, 39–40, 74, 79, 87, 90, 95, 96, 97, 104, 163
Iran, 37

Japan Steel Works (JSW), 58
Japan, 13, 39, 69, 78, 117, 131
    Fukushima accident (2011), xi, 12, 39, 41, 42, 48–50, 51, 52

Kazakhstan, 116
Kewaunee nuclear plant (Wisconsin), 70
Klein, Dale E., 108

La Hague, France, 131
left's energy policies, 4
light-water reactors (LWRs), 15, 103–04, 159
linear no-threshold (LNT) model, 35, 148–51
Lithuania, 48
loan guarantees, 88
Loris, Nicolas, 96
Louisiana Energy Services, 91
low enriched uranium (LEU), 117, 122, 162, 163
Loyola, Mario, 87

markets, benefit from, 96–98
Maryland, 71
Megatons to Megawatts Program, 121
Mercer County Community College, 92
methane, 24
Metropolis nuclear plant (Illinois), 117
Microsoft, 155
Midcontinent Independent System Operator (MISO), 4
mining and milling (uranium), 116
myths (nuclear power), 31–43
    global warming, 31–32
    not economically viable, 39–40
    nuclear accidents, 40–52
    nuclear reactors and terrorist attack, 35–36
    nuclear waste, 32–33
    nuclear weapons proliferation, 36–37
    radioactive emissions, 33–35, 34f
    transportation of nuclear materials, 37–39

Namibia, 112
National Enrichment Facility (New Mexico), 91

National Environmental Policy Act (NEPA) (1969), 71
National Highway Traffic Safety Administration (NHTSA), 63–64
natural gas plants, 22
natural gas prices, 101
natural gas reserves, 6, 10
Naval Nuclear Propulsion Program, 53
Netherlands, 162
net-zero carbon dioxide emissions, 8, 9, 10, 14, 80
North American Electric Reliability Corporation (NERC), 4, 8
North Anna nuclear station (Virginia), 42–43
North Carolina State University's College of Engineering, 92
North Korea, 37
nuclear accidents, 12–13, 39, 40–43
    policymakers respond to, 50–52
    regulatory system, 52–54
nuclear capacity, doubling, 15–17
nuclear economics, xii, 39–40, 71–75
nuclear energy
    as clean energy, 11–12
    density of, 11–12, 20–21, 20f
    mandates, 62–64
    nuclear fuel prices, 21–22, 21f
    politics and, xi– xii, 1
Nuclear Energy Innovation and Modernization Act (NEIMA), 78, 108
Nuclear Energy Institute, 91
Nuclear Engineering Enrollments and Degrees Survey, 91
nuclear fuel industry, *see* plutonium; thorium; uranium
nuclear fuel markets, fixing, 162–63
nuclear fuel policy reform, market principles to, 161–62
nuclear license applications, 69
nuclear plants
    closure of, 69–71, 70f
    construction costs, 19, 55–62

construction of (US experience), 56–62, 60*f*
costs of financing, 20
energy density, 11–12, 20–21, 20*f*
energy source, 21–23, 21*f*, 25–26
investment, xii–xiii, 58, 84–85, 91–92
levelized cost of electricity, 23–24, 24*f*
nuclear fuel supply chain, 23, 27
operating life of, 12
perpetual mediocrity loop, 64–67
prematurely retired reactors, 70
production cost of energy, 23
nuclear power, 12–13, 39, 40–43, 52–54
    broaden state authority to regulate, 151–53
    CO_2 reductions, 14–17
    global supply, 15–16
    nuclear waste, 13
    overregulation and costs of construction, 75–78, 76*f*
    radiation, 12–13
    subsidies, 25, 40, 61, 62, 82–83
    trade-offs, 12–14
Nuclear Regulatory Commission (NRC), 15, 20, 34, 36, 37, 42, 45, 91, 124, 127, 128, 148, 152–53
    licensing (SMR), 108
    linear no-threshold (LNT) model, 35, 148–51
    radioactive materials, transportation of, 38
    regulatory burden, transferring, 153–57
    regulatory mandates, 49, 128
    regulatory system, 52–54
    resident inspector program, 46
    Standards for Protection Against Radiation, 150
    Yucca Mountain program, 143
"nuclear renaissance," term, 145
nuclear waste, 13, 130–44
    recycling, 76–77, 130–31
    transportation of, 37–38
    used-fuel management, 134–38

Nuclear Waste Fee, 139, 140, 142
Nuclear Waste Fund, 132, 141, 142
nuclear waste management, 13, 27
    financing, 140–42
    fundamental problems, 135–36
    nuclear fuel policy reform, market principles to, 161–62
    transferring responsibility, 138–40
Nuclear Waste Policy Act (1982), 76, 109, 132–33, 142, 162
nuclear weapons, 29, 36–37
NuScale Power Corporation, 104
NuScale SMR, 105

Obama, Barack, 136
    administration, 57, 109
Olkiluoto 3 reactor (Finland), 55
Organisation for Economic Co-operation and Development (OECD), 22
Oyster Creek 1 nuclear plant (New Jersey), 70

Paducah enrichment plant (Kentucky), 121
Pakistan, 37
Palisades Energy, 155
Palisades nuclear plant (Michigan), 70
Paris Accords, 4
Peach Bottom high-temperature gas-cooled (HTGC) reactor, 103
Peach Bottom reactor (Pennsylvania), 125
per capita GDP, 2
Piketon nuclear plant (Ohio), 91
Pilgrim 1 nuclear plant (Massachusetts), 70
plutonium, 36–37, 130
policy revolution, xiii–xiv, 145–63
    broaden state authority to regulate, 151–53
    energy subsidies, ending, 146–48
    linear no-threshold (LNT) model, 35, 148–51
    NRC regulatory burden, transferring, 153–57
    nuclear fuel markets, fixing, 162–63

nuclear fuel policy reform, market
principles to, 161–62
private liability insurance
requirements, 157–61
policymakers, 7–8, 11
Pong (video game), 94
post-9/11, 35
POWER (magazine), 23
Presidential Directive 8 (PD-8), 78
pressurized water reactors (PWRs), 103
Price-Anderson Act (1957), 157–60
private liability insurance requirements, 157–61
Progress Energy, 92
Prohibiting Russian Uranium Imports Act (2024), 122, 126, 162–63
Public Service Enterprise Group, 92
public–private partnerships, 120
Purdue University, 92
Putin, Vladimir, 118, 119

radiation, 12–13, 45
linear no-threshold (LNT) model, 35, 148–51
regulatory reforms, inadequacy, 78–80, 89
renewable energy movement, 86–87
renewable portfolio standard (RPS), 60, 62, 85–86, 86*f*
Republic of Sakha, Russia, 110
Republicans, xi
Rickover, Admiral Hyman G., 105–09
Rosatom, 116, 117, 121
Russia, 27, 48, 78
advanced reactor market, 102, 110, 111
energy as a weapon, 119–21
energy imports, 118
enrichment capacity, 162
gas imports, 9, 52, 118
nuclear energy and national security, 28, 29, 30
nuclear fuel, recycling, 77
Ukraine, invasion of, 8, 9, 52, 111, 118
uranium/uranium trade, 112, 113, 126–29, 162–63

San Onofre 2 nuclear plant (California), 70
San Onofre 3 nuclear plant (California), 70
Seabrook nuclear plant (New Hampshire), 71
separative work units (SWUs), 117, 124
Shaw Group, 90
Shellenberger, Michael, 13
Shippingport nuclear plant (Pennsylvania), 15, 125
small- and medium-sized businesses, 8
small modular reactors (SMRs), 15, 99–111, 154, 155
applications, 100–01
commercial interest, 104, 105
designs, 100, 101–05
feasibility of, 102
government intervention, 108, 109
in the past, 102–05
licensing, 108
nuclear waste management, 108, 109
opportunity for, 109–11
partnerships, 100
Rickover's paradigm, 105–09
solar power, 25, 26, 27, 58–60, 61, 62, 95
"duck tail curve," 59–60, 60*f*
subsidies, 86, 87
South Carolina reactors, 77–78
South Korea, 69, 78, 111, 117
nuclear fuel, recycling, 33
and United States nuclear relationship, 28–30
spent nuclear fuel (SNF)
recycling, 76–77, 114
shipments, 37–38
used-fuel management, 134–38

Talen Energy, 155–56
Texas A&M, 92
thorium, 114, 125–26, 127, 151
Three Mile Island accident (1979), xi, 12, 41, 42, 44–46, 51, 132
Three Mile Island Unit 1 (Pennsylvania), 70, 155

Toshiba, 100, 104
transparency, 45–46
transportation, 32
Trump, Donald, 108
tsunami (Japan, 2011), 48, 49
Tubb, Katie, 96
Turkey Point Nuclear Units 3 and 4 (Florida), 155

U.S. News & World Report, 90
Ukraine, 48
    invasion of, 8, 9, 52, 111, 118
United Arab Emirates, 28
United Kingdom, 113, 124, 162
United Nations (UN), 13
United States International Trade Commission, 26
universities, 91–92
University of Florida, 92
University of Virginia, 92
uranium, 20, 36–37, 151
    cost of, 21–22
    deposits, 6, 114
    enrichment, 112–13, 117
    fissionable uranium, 130
    prices, 113–14, 114*f*
    production, 112–15
    recycling, 130–31
    Russian imports, ban on, 126–29
    supply chain, 116–19
    uranium fuel industry, fall of, 123–24
    US enrichment, 113, 120–21
uranium-233 (U-233), 125–26, 127
Urenco enrichment plant (New Mexico), 117, 121
USEC (United States Enrichment Corporation) (now Centrus), 91, 120
Uzbekistan, 110

Vermont Yankee nuclear plant (Vermont), 42, 70
video gaming company, 94–95
Virginia, 100, 114

Vogtle AP1000 reactors (Georgia), 15, 23, 27, 28, 55, 56–57, 77, 156

Watts Bar 2 reactor (Tennessee), 56
West Germany, 69
West Virginia, 100
Western Troy Capital Resources, 104
Westinghouse, 90–91, 104
wind power, 25–26, 27, 58–59, 62, 95
    subsidies, 86, 87
Wolverine Power Cooperative, 155
World Association of Nuclear Operators, 48
World Health Organization (WHO), 39, 41–42, 46–47, 50
World Nuclear Association, 27, 113

Yucca Mountain repository (Nevada), 33, 76, 77, 109, 131, 132, 135, 137, 142–44
    limitations, 133

Zaporizhzhia nuclear power plant (Ukraine), 36, 118
Zimmer nuclear plant (Ohio), 78

JACK SPENCER is a senior research fellow for energy and environmental policy, specializing in domestic and international nuclear energy policy, in the Center for Energy, Climate, and Environment at the Heritage Foundation. Spencer publishes on numerous issues relating to nuclear energy and is Heritage's go-to expert on nuclear waste management, technological advances, nuclear fuel, industry subsidies, and international approaches to nuclear energy. He also hosts the weekly Heritage Foundation–produced podcast *The Power Hour*, where he interviews top guests on energy and environment issues. Spencer has testified before Congress on numerous occasions, including on nuclear nonproliferation, nuclear waste management, and the role of nuclear in America's energy mix. He also testified before the Blue Ribbon Commission on America's Nuclear Future on how to solve the country's nuclear waste dilemma. Spencer oversaw the production of *Powering America*, a documentary film about the commercial nuclear energy industry, and served as a member of the Department of Energy landing team, working with career DOE staff to develop a departmental guide for incoming political leadership. Previously, Spencer oversaw research on a wide range of domestic economic and trade issues as vice president of the Institute for Economic Freedom at the Heritage Foundation. Prior to that, he served as director of the Roe Institute, where he spearheaded research initiatives on federal spending, taxes, regulation, energy, and the environment. He also served as a senior legislative analyst for Babcock & Wilcox, providing research, analysis, communications strategy, and outreach support for commercial nuclear power and navy nuclear programs.

Printed in the USA
CPSIA information can be obtained
at www.ICGtesting.com
LVHW010035231124
797380LV00002B/376